Lecture Notes in Electrical Engineering

Volume 611

The book series *Lecture Notes in Electrical Engineering* (LNEE) publishes the latest developments in Electrical Engineering - quickly, informally and in high quality. While original research reported in proceedings and monographs has traditionally formed the core of LNEE, we also encourage authors to submit books devoted to supporting student education and professional training in the various fields and applications areas of electrical engineering. The series cover classical and emerging topics concerning:

- Communication Engineering, Information Theory and Networks
- Electronics Engineering and Microelectronics
- Signal, Image and Speech Processing
- Wireless and Mobile Communication
- Circuits and Systems
- Energy Systems, Power Electronics and Electrical Machines
- Electro-optical Engineering
- Instrumentation Engineering
- Avionics Engineering
- Control Systems
- Internet-of-Things and Cybersecurity
- Biomedical Devices, MEMS and NEMS

For general information about this book series, comments or suggestions, please contact leontina. dicecco@springer.com.

To submit a proposal or request further information, please contact the Publishing Editor in your country:

China
Jasmine Dou, Associate Editor (jasmine.dou@springer.com)

India
Swati Meherishi, Executive Editor (swati.meherishi@springer.com)
Aninda Bose, Senior Editor (aninda.bose@springer.com)

Japan
Takeyuki Yonezawa, Editorial Director (takeyuki.yonezawa@springer.com)

South Korea
Smith (Ahram) Chae, Editor (smith.chae@springer.com)

Southeast Asia
Ramesh Nath Premnath, Editor (ramesh.premnath@springer.com)

USA, Canada:
Michael Luby, Senior Editor (michael.luby@springer.com)

All other Countries:
Leontina Di Cecco, Senior Editor (leontina.dicecco@springer.com)
Christoph Baumann, Executive Editor (christoph.baumann@springer.com)

** Indexing: The books of this series are submitted to ISI Proceedings, EI-Compendex, SCOPUS, MetaPress, Web of Science and Springerlink **

More information about this series at http://www.springer.com/series/7818

Tom J. Kazmierski • Sebastian Steinhorst
Daniel Große

Editors

Languages, Design Methods, and Tools for Electronic System Design

Selected Contributions from FDL 2018

 Springer

Editors
Tom J. Kazmierski
University of Southampton
Southampton, UK

Sebastian Steinhorst
Technical University of Munich
München, Bayern, Germany

Daniel Große
University of Bremen and DFKI GmbH
Bremen, Bremen, Germany

ISSN 1876-1100   ISSN 1876-1119   (electronic)
Lecture Notes in Electrical Engineering
ISBN 978-3-030-31587-0   ISBN 978-3-030-31585-6   (eBook)
https://doi.org/10.1007/978-3-030-31585-6

This Springer imprint is published by the registered company Springer Nature Switzerland AG.
The registered company address is: Gewerbestrasse 11, 6330 Cham, Switzerland

Preface

The increasing integration and complexity of electronic system design requires a constant evolution of the languages used and its associated design methods and tools. The *Forum on Specification & Design Languages* (FDL) is a well-established international forum devoted to the dissemination of research results, practical experiences, and new ideas in the application of specification, design, and verification languages to the design, modeling, and verification of integrated circuits, complex hardware/software embedded systems, and mixed-technology systems.

FDL is the main platform to present and discuss new trends as well as recent works in this domain. FDL 2018 was the twenty-first edition of the Forum and was held in September 2018 in Munich, Germany. Thanks to the commitment of the authors, presenters, and panelists, FDL 2018 was an extremely interesting and lively event.

This book contains a selection of papers which were presented at FDL 2018 and considered to be the best by both members of the program committee and participants of the Forum. The selection reflects the wide range of topics that were covered at this event.

The selected contributions particularly highlight that new modeling, verification, and implementation methodologies continue to extend the scope beyond advancing the established SystemC paradigm towards symbolic simulation, synchronous programming, and virtual prototyping, opening new cyber-physical application areas such as microfluidic devices and the Internet of Things.

By this, the portfolio of papers in this book provides an in-depth view of the current developments in our domain, which surely will have a significant impact in the future.

We would like to thank all the authors for their contributions as well as the members of the organizing and program committees and the external reviewers for their hard work in evaluating the submissions.

Special thanks go to Franco Fummi and his team from the University of Verona, who, together with the local team from Technical University of Munich, were responsible for the splendid organization of FDL 2018.

Finally, we would like to thank Springer for making this book possible.

Southampton, UK   Tom J. Kazmierski
Bayern, Germany   Sebastian Steinhorst
Bremen, Germany   Daniel Große
May 2019

Contents

Time in SCCharts

**Alexander Schulz-Rosengarten, Reinhard von Hanxleden,
Frédéric Mallet, Robert de Simone, and Julien Deantoni**

1 Introduction

Cyber-physical/embedded systems are typically *reactive*, meaning that they have to continuously react to their environment, and that these reactions must meet certain timing constraints. Real-time aspects may be rather simple, such as "the system must run at least at 10 KHz," or it may be quite intricate, like "coil A must be activated 27.3 ms after coil B." A long-standing challenge in the design of such *real-time systems* is to reconcile concurrency and determinacy. As it turns out, time plays a rather adversarial role there in that standard mechanisms to handle concurrency, such as Java/POSIX *threads*, are rather sensitive to how long individual computations take; determinacy is easily compromised by *race conditions* [18]. Synchronous languages address this challenge by abstracting from execution time; their semantics rests on the assumption that computations take zero time, and that outputs are synchronous with their inputs [5]. The synchronous programming paradigm has been explored since the 1980s, and, for example, SCADE (Safety-Critical Application Development Environment) and its certified code generator are routinely used for avionics control software [9].

The abstraction from time in synchronous languages typically comes at the price that all references to physical time must somehow be resolved by the environment. Unlike, for example, Harel's statecharts [13], which already included a mechanism to express timeouts, physical time is traditionally not a first-class citizen

A. Schulz-Rosengarten (✉) · R. von Hanxleden
Department of Computer Science, Kiel University, Kiel, Germany
e-mail: als@informatik.uni-kiel.de; rvh@informatik.uni-kiel.de

F. Mallet · R. de Simone · J. Deantoni
INRIA Sophia Antipolis Méditerranée, Sophia Antipolis Cedex, France
e-mail: frederic.mallet@inria.fr; robert.de_simone@inria.fr; julien.deantoni@inria.fr

© Springer Nature Switzerland AG 2020
T. J. Kazmierski et al. (eds.), *Languages, Design Methods, and Tools for Electronic System Design*, Lecture Notes in Electrical Engineering 611,
https://doi.org/10.1007/978-3-030-31585-6_1

in synchronous languages; they instead build on a *multiform notion of time*, where time is expressed by counting events (detailed further in Sect. 3.3). This is consistent with the synchronous abstraction, but in practice does not help the programmer much, who at the end of the day must express the required real-time behavior.

In this paper, we investigate how we can incorporate physical time into the synchronous model of computation. We do so using the SCCharts language [26]; however, the concepts presented here can be applied to other synchronous languages as well.

1.1   Contributions and Outline

- We show how timed automata, which model time with real-valued clocks, can be expressed in a synchronous setting with discretized execution (Sect. 2). Our proposal, which includes a new type **clock** for SCCharts, uses on-board mechanisms of synchronous languages (in particular *during actions*) to faithfully model clocks and imposes minimal requirements on the execution environment.
- We investigate the suitability of different execution regimes in a timed setting (Sect. 3) and argue that *dynamic ticks* [25] are a natural fit for realizing timed automata.
- We present an approach to implement dynamic ticks in a synchronous setting, where the compiler deduces anticipated tick durations from timing constraints in the model (Sect. 4).
- We propose a language extension of SCCharts (**period**) that allows to model multiclocked systems based on periodical activation of different subsystems and that maps naturally to real-valued clocks (Sect. 5).
- Finally, having clocks as first-class citizens we use them to map one abstract clock constraint, expressed in the Clock Constraint Specification Language (CCSL), to SCCharts (Sect. 6). This allows to not only relate the activation of subsystems to physical time, but also to the activation of other subsystems.

We briefly discuss further related work in Sect. 7 and conclude in Sect. 8.

2   Timed Automata in SCCharts

Timed automata, proposed by Alur and Dill [3], are a formalism to model the behavior of real-time systems over time. Timed automata consist of state-transition graphs with timing constraints using real-valued *clocks*. A timed automaton accepts *timed words*, which are (infinite) sequences in which a real-valued time of occurrence is associated with each element of the timed word.

Timed automata and their variations have been extensively studied for verification purposes [2, 3, 21]. We here want to use them for synthesis purposes as well.

That is, we investigate how to model the behavior of real-time systems such that the model can also be synthesized into a piece of software or hardware.

Timed automata have been extended in various ways, one example are *multirate timed automata* (or *multirate timed systems*) [2], where each clock has its own speed, possibly varying between a lower and an upper bound. Lee and Seshia [19] discuss (multirate) timed automata in the context of cyber-physical system design. One of their illustrating examples is the traffic light controller introduced in the next section.

2.1   The Traffic Light Controller Example

We use the traffic light controller shown in Fig. 1 as running example. The traffic light has three lights green, yellow, and red to control the car traffic and a button for a pedestrian to request secure crossing of the street, which should cause the traffic light to switch temporarily to a red light to stop the traffic. The automaton of the controller has a real-valued clock x, an input **pedestrian** indicating whether a pedestrian requests crossing the street, and three outputs sigR, sigG, sigY. The type **pure** denotes "pure signals" present or absent at each reaction and carrying no further data. The outputs do not directly indicate the light *states*, but rather constitute *events* that indicate color changes. It is assumed that initially the red light is turned on; emitting the event sigG switches off red and switches on green, and so on.

As shown in this example, a clock is represented by a first-order differential equation on a real number and can be explicitly set and used as transition guard. While in state red, time progresses with a slope of one ($\dot{x}(t) = 1$), this is also the case for all other states. Time is expressed in abstract time units; for simplicity, we assume for this example that one time unit corresponds to 1 s. Each transition has a

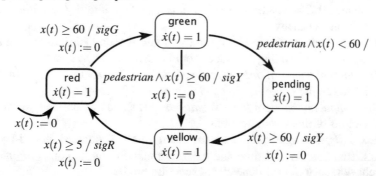

continuous variable: $x(t)$: $\mathbb{R}$
inputs: *pedestrian*: pure
outputs: *sigR, sigG, sigY*: pure

Fig. 1 Traffic light controller modeled as timed automaton. From Lee and Seshia [19] (CC BY-NC-ND 4.0)

guard, which consists of a *condition* (such as **pedestrian**) and a *timing constraint* (such as x ≥ 60), both of which are optional. When clock x has reached or surpassed 60, the automaton transitions to **green** emitting the green light and resetting the time to zero. Now the system waits for a pedestrian to push the button. When the **pedestrian** input is present, the reaction depends on the passed time. Case 1, if less than 60 s passed since entering **green**, the automaton will transition to **pending**, but x is not reset. It remains there until the time has reached at least 60, then the yellow light is turned on, the timer is reset and the state is switched to **yellow**. Case 2, if the **pedestrian** event occurs after at least 60 s in **green**, the automaton transitions directly to **yellow** with the same output and reset. After at least 5 s, the automaton leaves the **yellow** state for **red** and activates the red light and again resets the time. Note, the model could be simplified by omitting the direct transition from **green** to **yellow** and eliminate the second clause $(x(t) < 60)$ from the condition of the transition from **green** to **pending**. However, we leave the model as originally proposed [19].

2.2 Requirements for Time in SCCharts

To find a sound semantics for timed automata in SCCharts we first want to present the requirements that we consider sensible in this context.

Determinism The semantics should fit seamlessly into the synchronous paradigm and provide deterministic behavior, e.g., outputs are fully determined by inputs. For SCCharts that means there should be no changes to the underlying sequentially constructive model of computation.

Approximation of an eager semantics A solution must cope with runtime variations and imperfections of physical timers.

Scalability The number of (concurrent) timers should not be restricted.

Fine granularity The specification of time constrains should not be restricted by a specific granularity, e.g., one may have timeouts of 1 s and 3.1415926 ms in the same model. The resulting reaction time may be arbitrarily small but causes no performance penalty.

Time composability Time-based events should remain their intuitive semantics if composed, e.g., waiting 1 s twice should mean the same as waiting 2 s once.

Preservation of temporal order and simultaneity Timers that started in the same tick and run the same duration should expire in the same tick.

Minimal impact of physical timer variations It should be possible to avoid accumulations of timer imperfections.

Access to physical time and tick computation time The model should be allowed to access the physical time and tick computation time to compute time-related behavior, such as load-dependent execution modes.

Lean interface The inference between the model and its environment should be independent from the application, e.g., the interface should not change, if the number of timers changes.

Seamless compiler integration In the context of SCCharts any solution should fit into the Single Language-Driven Incremental Compilation (SLIC) concept and provide stand-alone features that are built on top of existing SCCharts, without the need to change the compilation back-end.

2.3 From Specification to Behavior: The Eager Semantics

Even though this traffic light controller specification seems rather clear and straightforward, it turns out that there is still some variation as to how the controller may behave in a specific scenario. The original definition of timed automata [3] is based on timed regular languages, where symbols in a word are associated with a real-valued time stamp. Formally, a *timed word* is a pair (σ, τ), where $\sigma = \sigma_1, \sigma_2, \ldots$ is an infinite word over some alphabet Σ of events, and a *timed sequence* $\tau = \tau_1, \tau_2, \ldots$ is an infinite sequence of time values $\tau_i \in \mathbb{R}$ that satisfies certain constraints (monotonicity and progress). Given a timed word, a *run* of a timed automaton is an (infinite) sequence of state transitions, analogous to standard regular languages defined by standard automata. For convenience, we extend the concept of timed words such that the inputs σ_i do not have to consist of exactly one event, but constitute arbitrary *input valuations* that assign a value and/or presence status to each input variable.

To make things concrete, assume that in our traffic light controller the pedestrian button is triggered at times 40 and 122.2. We denote this as input trace ($\langle$pedestrian, 40$\rangle$, $\langle$pedestrian, 122.2$\rangle$); we thus allow input sequences (timed words) to be finite, and we use a notation that associates each input valuation directly with a time stamp. Given such an input sequence, our timed automaton performs a sequence of *reactions*, or *ticks*, one for each time-stamped input valuation.

A first non-obvious question this raises is how system initialization should be handled. In principle, there is nothing that requires that the first reaction of the system must occur at time zero. Furthermore, the "initial transition" to state **red** is not really a transition, but rather a convenient way to specify initial values for variables, including clocks. However, it does seem reasonable to let the clock **x** assume the initial value 0 at time zero, and to make this explicit by performing an initial reaction with an empty input (denoted ϵ) at time zero. The resulting input trace is ($\langle\epsilon, 0\rangle$, $\langle$pedestrian, 40$\rangle$, $\langle$pedestrian, 122.2$\rangle$).

As illustrated in Fig. 2a, the traffic light controller reacts to this input trace by initializing itself at time 0, doing nothing at time 40, and then, at time 122.2, transitioning to **green** and emitting **sigG**. Then there is no further reaction due to the absence of further input events. However, this behavior is probably not what the creator of the traffic light controller intended. For example, the output **sigG** should probably not occur at time 122.2, even though 122.2 $\geq$ 60 certainly holds,

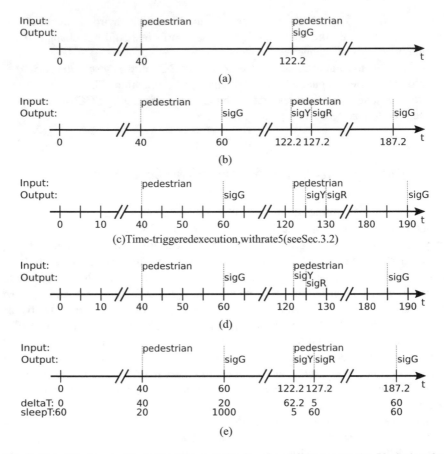

Fig. 2 Execution traces of the traffic light controller based on different semantics. Vertical strokes denote reactions. (**a**) Pure event-triggered execution (see Sect. 2.3/Sect. 3.1). (**b**) Eager semantics execution (see Sect. 2.3/Sect. 4.1). (**c**) Time-triggered execution, with rate 5 (see Sect. 3.2). (**d**) Time-event-triggered execution, multiform notion of time (see Sect. 3.3). (**e**) Dynamic tick execution (see Sect. 4.1)

but rather at time 60. Thus, we conclude that just the passage of time (without further input events) should also be able to trigger a reaction, in particular if the automaton contains transitions that are guarded solely by timing constraints. Lee and Seshia [19] resolve this by assuming that a transition is taken as soon as it is enabled; conceptually, the automaton reacts "continuously." This assumption, which we denote as *eager* semantics, leads to the trace in Fig. 2b, which augments Fig. 2a by further reactions, all with empty input valuations, at times 60 (emission of **sigG**, transition to **green**), 127.2 (emission of **sigR**), and 187.2 (**sigG** again). The remaining traces are explained in Sect. 3, along with their execution concepts.

2.4 Timed SCCharts

As it turns out, the synchronous model of execution fits quite naturally for timed automata as well. We here illustrate this with the SCCharts language. SCCharts provide many different language features; however, most of these are *extended features* that can be mapped to a very small set of *core features*. These extended features can be considered just as syntactical sugar, and the SCCharts compilation consists largely of model-to-model transformations that replace extended SCChart features by simpler features [26].

As we illustrate now, the timed-automata clocks can be added as an *extended SCChart feature* [26] without too much difficulty. Figure 3a shows the SCChart realization of the traffic light controller. Despite some minor syntactical differences, the structure of the state machine itself and its transitions and their effects are the same as in Fig. 1. The new SCChart keyword clock here declares a clock x, which then, as in timed automata, can be set to arbitrary values and can be used to guard transitions. We here use the float[1] data type for clocks, other types would also be possible. In fact, there are good arguments for using integral types, e.g., to preserve the associative law for additions.

Figure 3b presents the compiled intermediate result of TimedTrafficLight, revealing its actual internal implementation and behavior. In comparison to the original model in Fig. 3a, x is now an ordinary floating point variable, and the SCChart has an additional input deltaT. The only obligation on the runtime environment is, at each tick, to set deltaT to the time passed since the last tick. Based on these time increments, the SCChart itself keeps track of the progression of clocks. Specifically, the progression of time for the clock x is represented by during actions in each state, which increase the clock x by deltaT multiplied by the slope, which we omit here since it is 1. A during executes its effect in every tick its state is active, *except* for the tick the state is entered; this is important since only the time passed inside the state should be considered. Note that x may instantaneously assume up to three different values within a tick: the value at the beginning of a tick, the incremented value computed by the during action, and the reset value when a transition is taken that resets x to zero. This is no problem under the sequentially constructive (SC) semantics of SCCharts [26]. Applying the same idea to classical, non-SC synchronous languages would be a bit more involved, but with, for example, SSA-like renamings a synchronous language such as Esterel can also support multiple values per tick [22, 23].

---

[1]In SCCharts float is an abstract floating point number without actual precision limitations. We consider the choice of an actual data type orthogonal to the general concept presented here.

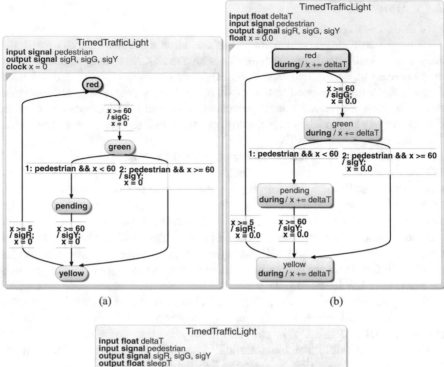

(a)   (b)

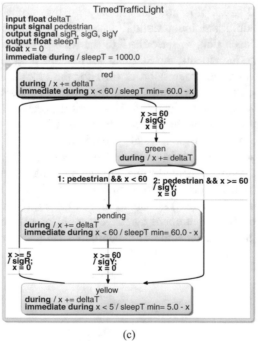

(c)

Fig. 3 Traffic light controller modeled as a timed automaton in SCCharts, with various compilation/expansion stages. (**a**) Original SCChart, with clock declarations. (**b**) Transformed SCChart. (**c**) Transformed SCChart with dynamic ticks

3   When to React?

Timed automata allow to add timing constraints to transitions based on a real-valued clock. It is clear that if the constraint is not met, the transition must not be taken. When the constraint *is* satisfied, the automaton *can* react. As discussed in Sect. 2.1, it seems advisable to tighten this by saying that we want to react as soon as possible, which we denoted as the eager semantics. Still, the non-trivial question remains of how to make sure in practice that reactions occur on time to implement an eager semantics, or, in practice, how to at least approximate it in some reasonable manner.

As it turns out, the question of when an automaton should react is not restricted to the "timed" setting presented here, but arises in synchronous programming in general. There, time is separated into discrete instants at which the system performs a reaction (*tick*). The *synchrony hypothesis* states that the reaction itself takes conceptually no time and that the actual time passes *between* reactions. Figure 4a illustrates this concept.

In practice, a synchronous program is synthesized into a tick function to perform reactions. Executing a tick takes computation time and separates inputs from outputs, as shown in Fig. 4b.

3.1   Event-Triggered Execution

In an entirely event-triggered execution, a reaction is triggered when an input (signal) changes. Hence our traffic light example would only react if the **pedestrian** input event occurs, as already illustrated in the trace in Fig. 2a. This execution regime is obviously insufficient as, e.g., it does not trigger transitions with only timing constraints, as discussed in Sect. 2.3. Consequently, a concept is needed which performs reactions based on time while handling the continuous nature of time.

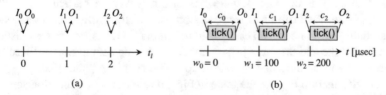

Fig. 4 Different timing abstractions [25]. (**a**) Logical time: time is discretized into logical ticks 0, 1, etc. Input I_i is synchronous with output O_i, the reaction time is abstracted to be 0. (**b**) Physical time: the computation of the i-th reaction, corresponding to logical tick i and the i-th call of the tick function, begins at *wake-up time* w_i. Inputs are read at the beginning of the computation, outputs are written at the end of the computation

3.2   Time-Triggered Execution

A common alternative to event-triggered execution is a periodical invocation of the tick function. One fixed global period is determined by analyzing the timing constraints of the model and its environment (i.e., poll rate of sensors), and sometimes also its worst case reaction time, to allow on-time executions of ticks. Figure 2c illustrates a trace with this execution semantics for our example in Fig. 3a. The period is 5, which is the greatest common divisor of the two relevant timing constants 5 and 60 in the model, and hence a sufficient sample rate for the systems timing constraints *if* events are discretized to this rate as well. The system only reacts every 5 s, which causes the **pedestrian** input occurring at time 122.2 to be processed *only* in the next period at time 125, consequently the **sigR** signal is also emitted at time 130. This behavior might be sufficient, especially when there are corresponding hardware sample rates for hardware sensors such as the pedestrian button.

Drawbacks of this execution regime are (1) the discretization of events (the **pedestrian** event is processed 2.8 s after its occurrence) and (2) efficiency. For example, for a delay of 60 as in **red**, there are always 12 ticks executed, even though the transition can only be taken in the 12th tick. The previous invocations are wasted processor time and energy, which is problematic especially in embedded use-cases.

3.3   The Multiform Notion of Time

When modeling temporal behavior, classical synchronous languages, such as Esterel, consider time as an arbitrary discrete input event to the program. For example, this could be a signal that is present in each tick a second has passed; however, equivalently, one could choose a signal that represents that a traveled distance has increased by one meter. The progression of time is measured by counting occurrences of some signal. This is also referred to as the *multiform notion* of time. This concept is quite flexible; however, in particular if multiple input signals are used to model time, say one signal for milliseconds and one for microseconds, this concept can easily lead to temporal inconsistencies, as discussed further by Bourke and Sowmya [8]; e.g., waiting for the next event "millisec" does not necessarily mean the same as waiting for 1000 events of "microsec."

For our SCChart in Fig. 3a, the trace in Fig. 2d represents an execution semantics using discrete timing events in combination with input event triggering. Since the model has two timing-related guards, 5 in state **yellow** and 60 in the others, we again opt for the greatest common divisor and use a timing event, let us denote it as **fivesec**, that indicates that 5 s has passed since the last occurrence of **fivesec**. As the trace illustrates, the system reacts every 5 s, always with **fivesec** present, and additionally at time 122.2 s, when **pedestrian** is present, but **fivesec** is absent. We call this *time-event-triggered* execution, since a reaction is triggered when either the timing event **fivesec** or some other event occurs.

Consider time 122.2, when the **pedestrian** input is processed and **sigY** is emitted. Since time is measured by counting **fivesec** events, and the last such event has occurred at time 120, the **pedestrian** event is effectively considered to have taken place at time 120. Consequently, **sigR** is *already* emitted at time 125 instead of 127.2; thus not 5 s has passed since **sigY**, but only 2.8 s, which is not compliant with the original traffic controller specification. Similarly, **sigG** is emitted at time 185, which is also earlier than in the trace in Fig. 2b. For this input trace, one could comply with the eager semantics by increasing the granularity of the discrete time event, i.e., using an event for 0.1 s passed. However, this would increase the number of reactions and load on the system significantly, while most of the reactions would not actually affect the state of the automaton.

3.4 Dynamic Ticks

To circumvent the difficulties of the execution regimes discussed so far, we here propose to not discretize time beforehand and to not model time by counting events, but propose to model time as continuous entity. Note that we still perform discrete reactions, only the time stamps are chosen from a real-valued domain, and in practice, this domain is also approximated by discrete types such as **float** (or **int** see Sect. 2.4).

This view of time as a continuous entity can be naturally combined with the concept of *dynamic ticks* [25], where the program itself outputs a request for how long the environment can wait or *sleep* until the tick function should be executed again, the *wake-up* time. Dynamic ticks can be combined with event-triggered execution; thus, one may again react to both the passage of time and external events. Note that this concept preserves the determinism of the synchronous system [25].

This results in a dynamic and efficient execution, as illustrated in Fig. 4b. The wake-up time w can either be set by an external global period or with dynamic ticks by the preceding tick function, adapting to the actually enabled reactions. Additionally, in situations where the reaction of the system depends on input events rather than time, dynamic ticks should be combined with event-triggered ticks, since no definite wake-up time can be determined. This is the case for the **green** state of the traffic light controller where the **pedestrian** input primarily triggers the transitions and the time constraints only separate which transition is taken. As discussed in the next section, dynamic ticks in combination with event-triggered execution allow the implementation of the eager semantics (trace in Fig. 2b).

4   Dynamic Ticks in SCCharts

Figure 5 illustrates the general structure that we propose to incorporate physical time into a reactive execution setting. As usual for an embedded system, a **Tick Function** communicates with its **Environment**, reading inputs from **Sensors** and conveying outputs to **Actuators**. Additionally, there is a **Trigger Unit** that calls the tick function, i.e., triggers one reaction (a tick). This classical setup is augmented by dynamic ticks, highlighted in red. Not only inputs trigger the execution (event-triggered) but there is also a **Time Manager** for time-triggered execution. This **Time Manager** is responsible for providing **deltaT**, the time passed since the last execution of the tick function, and it performs the waiting for the next time trigger based on **sleepT**. The new input and output extend the environment of the tick function. Note that this structure is still fully within the standard synchronous execution model, where the execution of a reactive system is divided into logical ticks, and a tick function reads certain inputs and produces certain outputs. Conceptually, **deltaT** is an input like any other input, and **sleepT** is an output like any other output. We also uphold our general requirement of determinacy: given a trace of inputs (including **deltaT**), the output trace (including **sleepT**) is fully determined.

4.1   The Traffic Light Controller with Dynamic Ticks

Our SCCharts traffic light control example in Fig. 3b can easily be further extended to use dynamic ticks, resulting in the SCChart shown in Fig. 3c. It has an additional output **sleepT** for the time span until the next time-related wake-up. In the root state there is an additional **immediate during** action, which executes its effect at every tick the state is active, including the tick the state is entered, due to the **immediate** modifier. It sets **sleepT** to an appropriate, presumably very large default value (1000.0 in this example), which conceptually denotes that there is no active timeout. **sleepT** is then updated in the states requesting an earlier

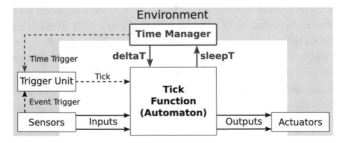

Fig. 5 Controller and environment of a dynamic tick function

wake up. This is done by further immediate during actions which register the remaining time until a guard of this state can be activated. The min= is an *update assignment* that assigns the minimal value between the current value of sleepT and the rhs expression. (As detailed further elsewhere [26], the SCChart semantics deterministically schedules "updates" such as +=, *=, etc. after other assignments; hence, there is no race condition between the assignment of the default sleep time and the min= assignments.) The requested sleep time is calculated from the timing constraints of the outgoing transitions, further discussed in Sect. 4.2.

The resulting behavior is illustrated in Fig. 2e. The dynamic reaction times emulate the eager semantics (Fig. 2b), which, as argued in Sect. 2.3, we chose as the preferable execution semantics for timed automata. The system reacts to the pedestrian input at time 40, the state of the automaton does not change; however, as illustrated by deltaT and sleepT presented under the time line, the dynamic ticks adapt to the event-triggered invocation and correctly compute a new sleep time of 20. After the output of sigG at time 60, no wake-up time can be computed since no transitions primarily depend on timing constraints, hence the default sleep time of 1000 is taken. The trace also shows that the reaction to the pedestrian event at 122.2 is also "on time," and the output of sigR is exactly 5 s after this event. Dynamic ticks use only a minimal number of reactions, to process all events and to perform all transitions at their expected time.

4.2 How to Compute Sleep Times

The main task in computing the sleep time is to detect if and which passage of time causes a transition to be enabled. Our SCChart compiler computes sleep times based on a static analysis of the timing bounds in the outgoing transitions of a state, with certain restrictions of timing constraint specifications to facilitate their implementation. More specifically, we look for timing constraints of the form $c \geq ltb$, where c is a clock and ltb some expression that we refer to as *lower timing bound*. We compute the corresponding sleep time as the difference between ltb and the current clock value. For example, state red in Fig. 3c has an outgoing transition with guard $x \geq 60$; hence, red gets augmented with an immediate during action that computes sleepT min = 60.0 - x. If a state has multiple outgoing transitions with lower timing bounds, we assign the minimal positive sleep time. To simplify the detection of lower timing bounds, our implementation rules out negations of timing constraints, but that does not limit expressiveness; for example, $!(x < 10)$ should be written as $x \geq 10$. Furthermore, constraints specifying an *upper* bound do not contribute to the sleep time since they, considered separately, do not require time to pass to be enabled and hence would result in a sleep time of zero.

Our example in Fig. 3c shows another case where no sleep time is requested and the value of sleepT should fall back on the default value. In state green, both outgoing transitions primarily depend on the pedestrian input, and x only distinguishes the two paths. The two timing constraints are *non-triggering* in that just the passage of time does not make a difference in whether any outgoing transition is enabled or not. If **pedestrian** is false, we do not take any transition, and if it is true, we take a transition, no matter what time it is; the time indicated by x solely decides *which* transition we take.

To detect such non-triggering timing constraints, assume that the i-th outgoing transition of some state has a guard $G_i = C_i \wedge T_i$, where C_i is a condition that does not depend on time and T_i is a timing constraint. Assume that no guard is currently active, i.e., $\bigvee_i G_i =$ **false**. Furthermore, assume that T_1 specifies a lower timing bound ltb. This would usually require the computation of a corresponding sleep time, unless T_1 is non-triggering—which is the case if $\exists i$ such that C_1 implies C_i and $\neg T_1$ implies T_i (i.e., whenever the ltb has not been reached yet, T_i holds). In our implementation, we further simplify the conditions and assume that C_1 and C_i are the same Boolean guard, and T_1 and T_i are negations of each other. As it turns out, the guards on the outgoing transitions from **green** fulfill that criterion, taking **pedestrian && x >= 60** for C_1 and **pedestrian && x < 60** for C_2; thus, the compiler classifies 60 to be a non-triggering ltb and does not compute a sleep time for it.

The concept of computing sleep times based on lower bounds is closely tied to the eager semantics. With a perfectly eager execution, it would be sufficient to write x $\geq$ 60 as x $=$ 60, but considering a real-valued time and a realistic implementation with possible timer imperfections, the first option is more robust and thus preferable. Similarly, we prefer closed timing intervals as specified with $\geq$ over open intervals specified with $>$.

However, the behavior specified by timing constraints can change when using a semantics other than the presented eager one or if the reactions are delayed. If, for example, more time than the minimum of a specified lower bound passes, it is possible that other transitions get also enabled or disabled, which may change the expected behavior. Assume the example that a state is entered when at least 60 s passed (x $\geq$ 60) and is immediately (in the same tick) left when at most 80 s have passed (x $<$ 80), without any reset of the clock. With eager semantics, the state will be entered after a time of 60 and then left immediately. If the reaction is delayed or another execution semantics is applied and the system is able to react after a time of 80 for the first time, then the state is entered but can never be left, leaving the system in that state forever. One could argue such system is designed badly, but this is the reason why we prefer the eager semantics. Note that for dynamic ticks we only trigger reactions on transitions that require time to pass, such as $\geq$ constraints, but when a $<$ constraint is the only guard, then there is no wake up. Otherwise such constraints with delayed transitions would cause a sleep time of zero which contradicts the concept of a delayed reaction. Note that delayed transitions, in contrast to immediate transmissions, require the state machine to stay for at least one tick in the state before it can be left using a delayed transition. In the previous example the transition x $<$ 80 is not delayed and the state can be left in the same reaction as entered.

Nevertheless, we also want to discuss the loosening of the eager semantics based on dynamic ticks in the next section.

4.3   Hard vs. Soft Bounds: The Greedy Semantics

Dynamic ticks can be further extended to introduce *soft bounds*, leading to a *greedy* semantics that loosens the regime of the eager semantics. To motivate, consider the minimal SCCharts example in Fig. 6a. There are two regions **Fast** and **Slow**, each one uses a timed automaton to react. Assume that the time scale of this example is microseconds; thus, **Slow** should react every millisecond and **Fast** three times faster. Hence, starting at time zero, the third reaction of region **Fast** will be at time 999, leaving only 1 μs to invoke the reaction of **Slow**, which might be infeasible for the environment.

To circumvent such short sleep times, we introduce soft bounds. States with an outgoing transitions with a soft bound still compute their own sleep time, but speculate to possibly "piggyback" on a somewhat earlier reaction invoked by another state.

Specifically, the user may, in region **Slow**, replace the *hard bound* $x \geq 1000$ by the *soft bound* $x \geq 999 \, || \, x \geq 1000$, as illustrated in Fig. 6b. Our implementation detects this pattern, as presented in Fig. 6c, and requests a sleep time of 1000 for this state, as for the original hard bound specified with $x \geq 1000$; however, at runtime the transition may already be taken at time 999, thus subsuming the sleep time of 1000. This favors earlier reactions over late reactions, prevents very small sleep times, and possibly reduces the total number of reactions.

4.4   Hard vs. Soft Resets: Managing Time

As a result of the eager semantics and its implementation using dynamic ticks, which assumes that the system reacts immediately to the enabling of a transition, the question arises: *Can we react in time?* The greedy semantics already allows to specify a certain amount of slack in the sleep time, but here we want to focus on the problem of possible deviation between the requested wake-up time and the actual execution time and their effects of the system, relevant for both dynamic ticks and execution with a fixed period. Since the system contains clocks, it requires a time input from the **Time Manager**, in case of SCCharts **deltaT**, and there are different options to provide this input.

The first option is an artificial simulation, where the passed time is always the time that should have passed, in case of the dynamic ticks the sleep time or the fixed period if used. This encapsulates the system in a perfect world, where the execution is independent from the real physical time.

The other option is to pass the real time to the system. That has the effect that the deviation between the requested wake-up time and **deltaT** that is passed to the system depends on the environment. We favor this option as it allows the model to react to timing overruns, for example, by entering a "degraded mode." However, in this case timer imperfections affect the clocks in SCCharts. Executing a tick *before*

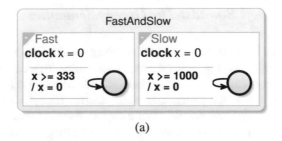

(a)

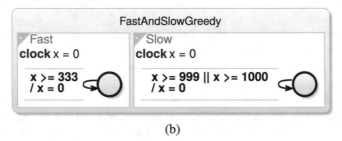

(b)

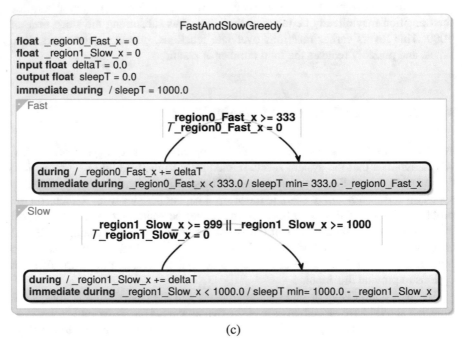

(c)

Fig. 6 Motivating example for using *soft bounds* in dynamic ticks. (**a**) SCChart motivating the use of soft bounds. (**b**) SCChart using soft bounds in the trigger in region Slow. (**c**) Transformed SCChart with soft bounds

Fig. 7 Variant of
FastAndSlowGreedy
SCChart using *soft reset* in
both regions

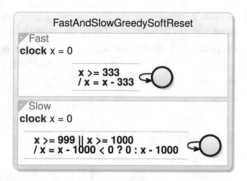

the requested sleep duration has passed does not affect the behavior since the clocks only increment by deltaT and the related time constraints do not trigger. When the tick is executed *after* the requested sleep time, the additional delay time will be present in all clocks.

We distinguish two approaches to handle this case, a *hard reset* sets the value of the clock to an absolute value, as presented in Fig. 6, and a *soft reset* that takes the additional delay into account. In Fig. 7 soft resets are used such that each region resets its clock x to the amount of time that exceeds the expected wake up ($x -$ 333 and $x - 1000$). Due to the soft bounds in region Slow, it is legal to take this transition with 999 s, this would result in negative clock value and thus the maximum of 0 and $x - 1000$ is assigned to x.

The consequence of hard resets is that clocks start to "drift" as soon as the tick function invokes a slower than the expected wake up. For example, if region **Fast** in Fig. 6 wakes up at 335 s, it would reset the clock to 0 and request a sleep time of 333 s, disregarding the 2 s that additionally passed. Hence the (earliest) next wake up would be at 668 s and this drift increases as the delays accumulate over time. This violates our requirement of temporal order and simultaneity; hence, we prefer a soft reset that resists the accumulation of timer imperfections. In the SCChart in Fig. 7 the **Fast** region would reset its clock to 2 if it woke up after 335 s and consequently would only request a sleep time of 331 s. The **Slow** region behaves similarly.

5 Multiclock SCCharts

Timed automata naturally support multiple clocks and so does its SCCharts implementation. In synchronous languages, there is also the concept of multiclocking [12], as in Multiclock Esterel by Berry and Sentovich [6]. In that context the term "clock" does not relate to a real-valued time measurement but a hardware clock that drives a hardware circuit or similarly designed software. In multiclock systems, different parts of the program are activated by different clocks, which are additional inputs to the program and effectively refine the base clock. Our concept presented so far can be further adapted to allow such multiclocking.

We have augmented SCCharts with an additional extended feature period, which controls the activation of states and regions based on a real-time clock. The period command ensures that the guarded state or region is only activated if the given amount of time has passed since the entering/start of the state/region or its last activation.

5.1   The Motor Example

To illustrate the usage and effect of the period feature, Fig. 8 presents the SCCharts example Motor. This represents a controller for two rather simplified stepper motors, for example, to drive a robot. There is a left (motorL) and right (motorR)

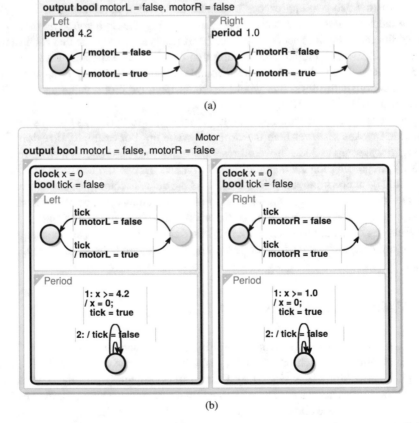

Fig. 8 Motor example modeled in SCCharts with periodic regions. (**a**) SCChart with period annotation. (**b**) Transformed SCChart

motor, which are run by toggling the corresponding Boolean output at a certain frequency. The SCChart has two concurrent regions, each controlling one motor with a simple state machine with two states. The transitions cycle between the states and toggle the motor variable. In our example, assuming time units of ms, the left motor must toggle every 4.2 ms, which is represented by the **period** annotation in the region. The right motor is run with a period of 1 ms.

To inspect the internal implementation of the extended **period** feature, Fig. 8b shows the compiled intermediate result of **Motor**. The **periods** are transformed into timed automata, as introduced in Sect. 2, to control the timing of the regions. In region **Left**, the inner states of the region are moved into a new super state that declares a new clock variable x and a Boolean flag **tick**. The tick variable acts as guard for all reactions in the original state machine, now present in the inner region named **Left**. This prevents the inner SCChart from performing any action if the clock is false. If any transition or action has its own guard, it would be conjuncted with **tick**. Here **tick** is initialized to false, which means that no reaction takes place in the initial tick; however, we might also initialize **tick** to true, which would cause an initial reaction.

There is also a new region **Period** with a single-state timed automaton. At each tick when the clock x reaches the period's threshold, the **tick** variable is set to true and enables the reaction in the other region. Otherwise, indicated by the transition with the lower priority (2:), the variable is set to false. Analogously, the **Right** region is affected by the period transformation. Note that one might also add the clock logic directly within the existing **Left** and **Right** regions. However, we decided to add the separate **Period** region and explicit **tick** guard, to reduce the number of timed guards in the model and to have a clear separation between timing and the original SCChart. In the process of compiling SCCharts, the next step would be to transform the **clock** feature as conceptually presented in Fig. 3.

6 Extension with Clock Patterns

As we have introduced clocks and tick flags that represent activation conditions of regions, we discuss here some possible use of those ticks. In particular, we want to make explicit relationships between these ticks just as in polychronous systems [17] and multiclock implementations [12]. The Clock Constraint Specification Language (CCSL) [4] has been defined as a language to handle clocks and to specify pure clock-related constraints independently of a specific programming language. CCSL sees clocks as infinite sequences of ticks and can define when a tick (therefore a region) should tick or cannot tick. We propose to annotate an SCChart with CCSL constraints that make explicit the rate relationships amongst the various regions and states. This can be done as a pure syntactic extension as long as such a specification can be compiled (internally) into a valid SCChart.

CCSL provides a concrete syntax to handle clocks, whether logical or physical, as first-class citizens. It provides patterns of classical clock constraints (like periodic, sporadic) that can be of three types: *synchronous* clocks are directly inspired from primitive constructs of synchronous languages [5]; *asynchronous* clocks rely on the relation "happens-before" from Lamport's logical clocks [16]; and *real-time* clocks represent physical time. Real-time constraints are usually a special case of the logical ones. For instance, CCSL defines both a real-time and logical notion of periodic behavior. A clock a is periodic on another clock b with period p if a ticks synchronously at every p^{th} tick of b. If b is a physical (real-time) clock (e.g., s), then it is a classical periodic behavior; otherwise, it remains purely logical. The semantics of each CCSL constraint is an automaton and a CCSL specification is the synchronized parallel composition of those automata [20].

Synchronous constraints are encoded as pure finite-state automata. *Asynchronous constraints* rely on state machines with unbounded integer counters. In TimeSquare [10], real-time constraints are encoded as a composition of logical constraints. However, real-valued clocks can also be encoded as timed automata [24], and the dynamic tick mechanism provides an efficient way to encode them in SCCharts. The goal here is to annotate an SCChart with CCSL constraints. This relies on the explicit tick flag introduced in Fig. 8b. CCSL annotations can either force the tick to occur (and therefore the region to execute) or observe unexpected behaviors and raise alarms. Both examples are illustrated in this section.

Figure 8a illustrated real-time clocks. In that model, the periodic behaviors of both regions are relative to an absolute real-time reference, assumed to be ms in that example. Alternatively, we can define the relative periodicity of the regions in CCSL as some rational period p, as in **repeat** left **every** 4 right, where left is a clock associated with the left region and right is a clock associated with the right region. The semantics of this constraint is given as a simple finite-state automaton that can be encoded as an SCChart in a straightforward way, as illustrated in Fig. 9. There the guard "4 right" is a *count delay* that becomes enabled after four occurrences of right.

Such synchronous constraints specify a fully determined behavior. When using asynchronous constraints, we may get a partially undetermined behavior. Consider, e.g., a periodic behavior with jitter as in **repeat** left **every** [4,5] right. This constraint expands as the following primitive CCSL constraints:

Fig. 9 Expansion of logical
periodic constraint

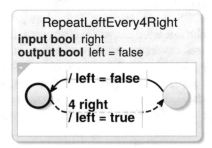

1 lower = pLeft **delayedBy** 4 right
2 upper = pLeft **delayedBy** 5 right
3 lower < left ≤ upper

where **pLeft** is defined by the constraint left = pLeft $ 1. The $ is used for unit-delay as in Signal, it defines **pLeft** as the same clock as left preceded by one more tick, like a **pre** operator.

lower represents the lower bound for **left** to tick, while **upper** is the upper bound. The last equation forces **left** to tick strictly after **lower** and before **upper**. Each of these constraints can be encoded as concurrent SCChart (see Fig. 10a). TimeSquare builds the synchronized product of these automata to compute a finite-state automaton that can be encoded as a simpler SCChart, see Fig. 10b; note that the dashed transition leaving **s0** is *immediate*, meaning that it could be taken immediately in the tick when **s0** is present.

The behavior exposed in Fig. 10 describes clock relations between the two regions **left** and **right**. At the same time, it observes whether or not the regions behave as expected. Clocks **left** and **right** are inputs and a wrong sequence of inputs would lead into the **error** state, like an assertion. It also enables or disables the code in regions. In state **enabled_left**, the region **left** is enabled and its code is executed as expected. In other states, the region is disabled and its code should be ignored.

7 Related Work

Timed automata [3] introduce real-valued clocks to describe the temporal behavior of systems using a continuous notion of time. Usually some progress conditions are required [14] to avoid time divergence and to guarantee that the system does not remain idle forever. While timed automata and their multiple extensions are originally defined with an asynchronous semantics, we here propose to harness them in the synchronous settings of SCCharts.

As presented here, the **clock** feature models single-rate clocks, as initially proposed by Alur and Dill [3], since it relieves the modeler of explicitly handling time. However, note that the **clock** type is only a convenience feature, and a user can always model SCCharts directly as presented in Fig. 3b and implement multirate clocks by scaling the change of **x** in the **during** actions.

Altisein and Tripakis [1] implement timed automata by wrapping them into a *global execution model* that captures the real time and handles the timers in the original model. This approach allows to keep the original semantics of timed automata and capture the influence of the execution platform on timers to enable verification. With **clocks** in SCCharts we do not address the topic of verification but follow the same approach, as we process the raw real time from the environment and allow the model to handle any timer imperfections, for example, by performing a *soft reset* (Sect. 4.4).

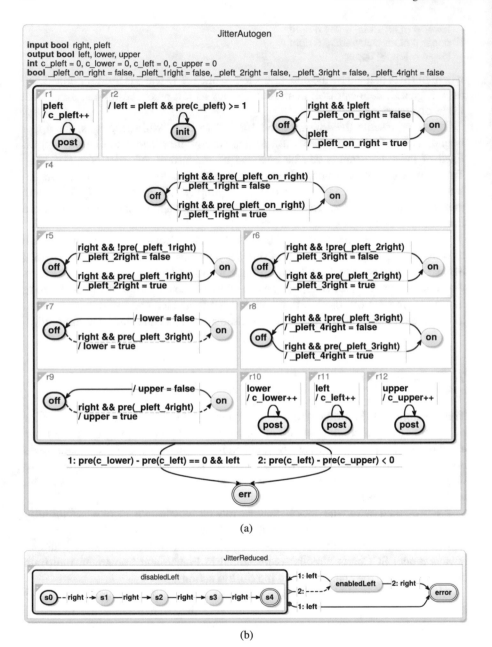

(a)

(b)

Fig. 10 Periodic behavior with jitter. (**a**) Automatically generated SCChart for jitter. (**b**) Simplified SCChart for jitter

As discussed by Sifakis et al., multirate timed automata can be mapped to timed automata [21], and in the traffic light example, that transformation is rather straightforward as the only clock x always moves at the same speed.

Harel [13] also proposes time extensions to statecharts where a time t is associated with every transition and t refers to a global notion of discrete time steps. We consider here both discrete and real-valued models of time.

Zelus [7] is a synchronous language that mixes both discrete-time and continuous-time behaviors. Continuous behaviors are described through ordinary differential equations. We are not describing continuous behaviors here but provide an extension to SCCharts to make explicit the activation conditions of regions under the form of clocks that serve to express both real-valued and logical constraints.

As explained, this work builds on the concept of dynamic ticks proposed by von Hanxleden et al. [25]. Thus most of the related work discussed by them is also relevant for this work. This includes, for example, the work by Jourdan et al. on extending ARGOS with timing constructs [15], or PTIDES (Programming Temporally Integrated Distributed Embedded Systems), which addresses the design and implementation of distributed real-time embedded systems [11].

The dynamic ticks are akin to the agenda of timed events used in discrete event system specification (DEVS) [27] to always pick the most urgent event without relying on a timed-triggered strategy. However, the sequentially constructive semantics of SCCharts, which permits instantaneous modifications of variables under consideration of data dependencies, reduces the need for the so-called delta-cycles.

8 Conclusions and Outlook

We have investigated how to incorporate physical time into the synchronous model of execution. As it turns out, timed automata can be mapped naturally to the synchronous setting, requiring only minimal support from the environment. However, to achieve a concrete implementation also requires to settle for a concrete, unambiguous semantics that specifies not only when a system *may* react but also when it actually *does* react; to that end, we have settled for the eager semantics, as also suggested by Lee and Seshia [19].

We have proposed two extensions to SCCharts, namely clocks and periods, that can be mapped directly to standard SCCharts. We have implemented these extensions as part of an open-source compiler.[2] We expect that similar extensions could be implemented in other synchronous languages, such as Esterel, Lustre, or also SCADE, as they, for example, also facilitate the "during actions" required for tracking clocks.

---

[2]http://rtsys.informatik.uni-kiel.de/kieler.

With this concept we fulfilled our own requirements, presented in Sect. 2.2, by providing a semantics that is based on the (synchronous) sequentially constructive model of computation and ensures deterministic behavior (Sect. 2.4). With dynamic ticks we implemented the eager semantics with a lean interface (Sect. 4) and allow timing constraints (clocks) with fine granularity, preserved simultaneity, and good composability and scalability. Soft resets allow to minimize the impact of physical timer variations (Sect. 4.4) and the implementation of clocks is seamlessly integrated into the SCCharts complier as an extended feature (Sect. 2.4).

We have cast the concept of clocks and time in the context of physical time and durations. However, for us the only practical requirements on clocks are the ones that timed automata cast on clocks, namely monotonicity and progress. Thus, one might also consider other (at least conceptually) continuous entities as clocks, such as distance traveled. In other words, the multiform notion of time could also be applied to time as proposed here, all within a synchronous setting.

There are several directions to proceed from here. First, we would like to get more practical experience with the language constructs proposed here. The clock and period extensions already promise to be quite useful, but other, more high-level language extensions would be feasible as well, as suggested, for example, by the features already present in CCSL. Then, while the way these features are mapped to standard SCCharts seems natural and straightforward, more efficient mappings might be possible. Similarly, in the context of dynamic ticks, we currently have a rather simple heuristics to compute sleep times from timing constraints; more powerful static analyses might again lead to a more efficient implementation.

References

1. Altisen, K., & Tripakis, S. (2005). Implementation of timed automata: An issue of semantics or modeling? In *Formal Modeling and Analysis of Timed Systems* (pp. 273–288). Berlin: Springer.
2. Alur, R., Courcoubetis, C., Halbwachs, N., Henzinger, T. A., Ho, P.-H., Nicollin, X., et al. (1995). The algorithmic analysis of hybrid systems. *Theoretical Computer Science, 138*(1), 3–34.
3. Alur, R., & Dill, D. L. (1994). A theory of timed automata. *Theoretical Computer Science, 126*, 183–235.
4. André, C. (2009). Syntax and semantics of the clock constraint specification language (CCSL). Research Report RR-6925, INRIA.
5. Benveniste, A., Caspi, P., Edwards, S. A., Halbwachs, N., Le Guernic, P., & de Simone, R. (2003, January). The synchronous languages twelve years later. In *Proceedings of the IEEE, Special Issue on Embedded Systems* (Vol. 91, pp. 64–83), Piscataway, NJ: IEEE.
6. Berry, G., & Sentovich, E. (2001). Multiclock Esterel. In *CHARME '01: Proceedings of the 11th IFIP WG 10.5 Advanced Research Working Conference on Correct Hardware Design and Verification Methods* (pp. 110–125), London: Springer.
7. Bourke, T., & Pouzet, M. (2013, April). Zélus: A synchronous language with odes. In *Proceedings of the 16th international Conference on Hybrid Systems: Computation and Control, HSCC 2013*, Philadelphia, PA (pp. 113–118).

8. Bourke, T., & Sowmya, A. (2009, November). Delays in Esterel. In *SYNCHRON'09— Proceedings of Dagstuhl Seminar 09481*, number 09481 in Dagstuhl Seminar Proceedings. Internationales Begegnungs- und Forschungszentrum (IBFI), Schloss Dagstuhl (pp. 22–27).

9. Colaço, J.-L., Pagano, B., & Pouzet, M. (2017, September). SCADE 6: A formal language for embedded critical software development (invited paper). In *11th International Symposium on Theoretical Aspects of Software Engineering TASE*, Sophia Antipolis (pp. 1–11).

10. Deantoni, J., & Mallet, F. (2012). Timesquare: Treat your models with logical time. In *50th International Conference on Objects, Models, Components, Patterns (TOOLS). Lecture Notes in Computer Science* (Vol. 7304, pp. 34–41). Berlin: Springer.

11. Eidson, J., Lee, E. A., Matic, S., Seshia, S., & Zou, J. (2012, January). Distributed real-time software for cyber-physical systems. *Proceedings of the IEEE, 100*(1), 45–59.

12. Gamatié, A., & Gautier, T. (2010). The Signal synchronous multiclock approach to the design of distributed embedded systems. *IEEE Transactions on Parallel and Distributed Systems, 21*(5), 641–657.

13. Harel, D. (1987, June). Statecharts: A visual formalism for complex systems. *Science of Computer Programming, 8*(3), 231–274.

14. Henzinger, T. A., Nicollin, X., Sifakis, J., & Yovine, S. (1994). Symbolic model checking for real-time systems. *Information and Computation, 111*(2), 193–244.

15. Jourdan, M., Maraninchi, F., & Olivero, A. (1993, June/July). Verifying quantitative real-time properties of synchronous programs. In *Proceedings of Computer Aided Verification (CAV'93). LNCS* (Vol. 697, pp. 347–358).

16. Lamport, L. (1978, July). Time, clocks, and the ordering of events in a distributed system. *Communications of the ACM, 21*(7), 558–565.

17. Le Guernic, P., Talpin, J.-P., & Le Lann, J.-C. (2003). POLYCHRONY for system design. *Journal of Circuits, Systems, and Computers, 12*(3), 261–304.

18. Lee, E. A. (2006). The problem with threads. *IEEE Computer, 39*(5), 33–42.

19. Lee, E. A., & Seshia, S. A. (2017). *Introduction to Embedded Systems, A Cyber-Physical Systems Approach* (2nd ed.). Cambridge: MIT Press.

20. Mallet, F., & de Simone, R. (2015). Correctness issues on MARTE/CCSL constraints. *Science of Computer Programming, 106*, 78–92.

21. Olivero, A., Sifakis, J., & Yovine, S. (1994). Using abstractions for the verification of linear hybrid systems. In *Proceedings of the 6th Annual Conference on Computer-Aided Verification, Lecture Notes in Computer Science 818* (pp. 81–94). Berlin: Springer.

22. Schulz-Rosengarten, A., Smyth, S., von Hanxleden, R., & Mendler, M. (2018, June). On reconciling concurrency, sequentiality and determinacy for reactive systems — A sequentially constructive circuit semantics for Esterel. In *2018 18th International Conference on Application of Concurrency to System Design (ACSD)* (pp. 95–104).

23. Schulz-Rosengarten, A., Smyth, S., von Hanxleden, R., & Mendler, M. (2018, February). A sequentially constructive circuit semantics for Esterel. Technical Report 1801, Christian-Albrechts-Universität zu Kiel, Department of Computer Science. ISSN 2192-6247.

24. Suryadevara, J., Seceleanu, C. C., Mallet, F., & Pettersson, P. (2013, September). Verifying MARTE/CCSL mode behaviors using UPPAAL. In *Software Engineering and Formal Methods. Lecture Notes in Computer Science* (Vol. 8137, pp. 1–15). Berlin: Springer.

25. von Hanxleden, R., Bourke, T., & Girault, A. (2017, September). Real-time ticks for synchronous programming. In *Proceedings of the Forum on Specification and Design Languages (FDL '17)*, Verona.

26. von Hanxleden, R., Duderstadt, B., Motika, C., Smyth, S., Mendler, M., Aguado, J., et al. (2014, June). SCCharts: Sequentially Constructive Statecharts for safety-critical applications. In *Proceedings of the ACM SIGPLAN Conference on Programming Language Design and Implementation (PLDI '14)*, Edinburgh (pp. 372–383). New York: ACM.

27. Zeigler, B. P. (1976). *Theory of Modeling and Simulation*. New York: Wiley.

Generation of Functional Mockup Units for Transactional Cyber-Physical Virtual Platforms

Stefano Centomo, Michele Lora, and Franco Fummi

1 Introduction

Cyber-Physical Systems (CPSs) are shaping today's world. They are an enabling technology for many different ongoing technological disruptions, such as smart manufacturing, autonomous driving, etc. As such, improving design methodologies for CPSs is crucial to advance a wide set of system engineering sub-disciplines [18].

System design requires models to be simulated providing designers with the feedback necessary to evaluate the quality of their ideas [12]. The heterogeneity of CPSs makes modeling and simulation pretty intricate tasks [13]. To achieve holistic simulation of such heterogeneous systems, designers must either rely on complex co-simulation environment aggregating specialized simulators for the many design domains involved in the system or produce a single holistic model of the system [16]. However, the latter solution requires to access, often unavailable, open specifications for every component of the system. On the other hand, co-simulation requires interfacing different simulation tools. Such tools often provide incompatible interfaces, thus requiring time-consuming adapters [11].

In this scenario, the FMI standard for co-simulation emerged as one of the most promising technologies to interface heterogeneous simulators [1]. It defines an Application Programming Interface (API) that must be implemented by the

S. Centomo (✉) · F. Fummi
Department of Computer Science, University of Verona, Verona, Italy
e-mail: stefano.centomo@univr.it; franco.fummi@univr.it

M. Lora
Singapore University of Technology and Design, Singapore, Singapore
e-mail: michele_lora@sutd.edu.sg

© Springer Nature Switzerland AG 2020
T. J. Kazmierski et al. (eds.), *Languages, Design Methods, and Tools for Electronic System Design*, Lecture Notes in Electrical Engineering 611,
https://doi.org/10.1007/978-3-030-31585-6_2

simulator. As such, FMI is well suited to build *Cyber-Physical Virtual Platforms* emulating both the "cyber" and "physical" parts of a CPS [15].

Even though the FMI standard proved to be a powerful tool to build such Cyber-Physical Virtual Platform, its focus is still strongly oriented to the simulation of continuous dynamic systems [19]. Thus, simulation of digital components still requires adapting the use of the standard to replicate the semantics of HW simulators [15]. Improvements to support Hardware Description Language (HDL) models in FMI have been addressed [8, 15]. However, the advantages in terms of simulation speed of higher-level models, such as *Transaction-level models* [6], have not been exploited so far due to some limitations of the standard. This chapter aims at analyzing and discussing such limitations. Then, it proposes a set of adjustments in the use of FMI constructs defined in the current standard for co-simulation (i.e., version 2.0). Furthermore, it presents a simulation coordination scheme that exploits such adjustments. These contributions together allow generating *Transaction-level FMUs* for Cyber-Physical Virtual Platforms.

In the last few months, the FMI Steering Committee announced a new interface (version 3.0) that aims to introduce the hybrid co-simulation concept [10]. However, it is still in alpha release and, as highlighted by the analysis presented in this chapter (Sect. 6), it still require many improvements to effectively enhance the support of digital components into FMI-based simulation environments. On top of the time that will be necessary to develop the new standard, any new version of the standard will also require time to be accepted from all the tools supporting the previous standard. Meanwhile, using the current version 2.0, as we do in the approach presented by this chapter, guarantees compatibility with the current version of the tools.

Figure 1 summarizes the contributions of this work. On the left, the CPS to be designed is simulated by using a *Cycle-accurate Cyber-Physical Virtual Platform*. The virtual platform is composed by exploiting the FMI standard. It is composed of both the models of the "cyber" and the "physical" sub-systems of the model. In this work we focus on the "cyber" part of the system modeled by aggregating different FMUs, each of them representing a digital components of the system. The simulation is managed by a *Master Algorithm* coordinating the FMUs. The time evolution of the virtual platform on the left side is accurate with respect to the clock cycle of the system: each simulation step simulates a single clock cycle, synchronizing at each step. This work improves the left side configuration by proposing two modifications to the platform and its components:

- The functionality within the **FMUs** composing the digital part of the system is abstracted to **transaction-level**. Their interfaces are modified to make them communicate their internal local time backward to the master algorithm.
- The **master algorithm** is improved to exploit the information about the local time of the FMUs in the model.

These modifications allow to produce the *Transaction-level accurate Cyber-Physical Virtual Platform* on the right-hand side of the figure. The platform synchronizes at each transaction defined by the communication protocol. Thus, it benefits the lighter synchronization for improving the simulation speed.

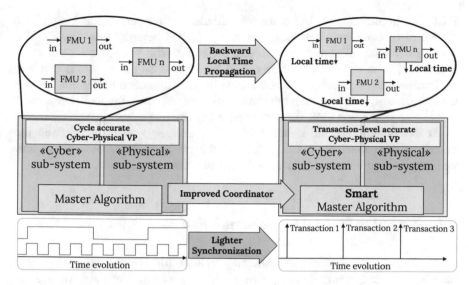

Fig. 1 Overview of the contribution

This chapter is organized as follows: Sect. 2 gives the necessary background about FMI, and summarizes the state of the art. Section 3 discusses the advantages and the limitations in the current version of the FMI standard and discusses a set of possible improvements. Section 4 presents the methodology proposed by this paper. The presented approach is implemented by building an automatic tool-chain and then experimentally evaluated in Sect. 5. Then, Sect. 6 updates the discussion we previously presented [9] about the current support for digital models within FMI-based simulation environment. It presents the current efforts being made by the FMI Steering Committee to develop a novel version of the standard, and it discusses the improvements necessary to improve the support of discrete models into hybrid systems. Finally, in Sect. 7 we draw some conclusions.

2 Background and Related Work

FMI is a tool-independent standard aiming to enhance the interoperability between tools of different vendors in the field of systems design [1, 17]. It supports both model exchange and co-simulation of dynamic models produced by using different tools and languages. The standard has been originally developed by Daimler AG, and maintained initially by the MODELISAR Consortium, and by The Modelica Association after the MODELISAR European Project ended. The latest version of the standard is the 2.0 of 2014. Currently, the version 3.0 is under development. The basic blocks of any FMI-based simulation environment are called FMUs. Multiple FMUs can be imported within a simulation tool to be executed. Each

FMU may implement only one of the two variations of the current standard: *Model Exchange* or *Co-Simulation*. Model exchange FMUs describe functionalities by using differential, algebraic, and discrete equations with time-, state-, and step-events [17]. The equations must be solved by an external solver that is required to simulate model exchange FMUs. Meanwhile, Co-simulation FMUs must model the functionality and implement the solver as well. As such, the model described within a co-simulation FMU does not require any external solver.

At its current state, the standard for model exchange does not suit well for describing discrete-event models [15]. Thus, this chapter focuses on co-simulation whose main features and structure are described hereby.

2.1 FMI Standard 2.0 for Co-simulation

Practically, an FMU is an archive containing an XML file describing the component interface and a dynamic library providing its implementation. Furthermore, the dynamic library contained in any FMU for co-simulation must implement also the solver necessary to execute the functionality. The XML file must specify all the variables of the FMU visible to the simulation environment [1]. Each variable is characterized by a *name*, *causality* (e.g., input, output, parameter, etc.), a *type*, and a *value reference*. The value reference of a variable must be unique among the variables of each type. Each variable is uniquely identified by the pair made of its type and value reference. The dynamic library must implement the functionality by implementing a set of functions defined by the standard. The most important, among the many defined in the current version of the standard, are:

- `fmi2SetupExperiment`: initializes the internal variables of the FMU.
- `fmi2Set`: sets the value of an internal variable of the FMU, i.e., it assigns a value to an input.
- `fmi2Get`: gets the value of an internal variable of the FMU, i.e., it returns the value of an output.
- `fmi2DoStep`: advances the simulation time of the component executing the behavior defined by the model.

The dynamic library must be generated using C-like linking [8], as such the functionality is usually expressed by using either C or C++. The standard defines the signature for all the C functions to be implemented by the dynamic library. However, it does not impose how they should be used, as it rather defines only some limitations on the possible combinations.

2.2 Simulation Coordination in the FMI Standard

Any model having one or more FMUs requires a coordination mechanism compliant with the FMI standard. Version 2.0 of the standard [1, 17] defines the concept of *master algorithm* as the module of managing communication and synchronization for sets of FMUs. Communication is managed by the master algorithm by invoking the `fmi2Get` and `fmi2Set` functions of the co-simulation API. Meanwhile, synchronization and simulation advancement is implemented by carrying on the components execution by invoking the `fmi2DoStep` functions of the FMUs composing the model being simulated. The standard defines some rules about how the master algorithm should be. However, the exact definition of the algorithm is not part of the standard. In fact, the rules explicitly defined are mostly imposing some limitations on the structure.

Figure 2 reports a statechart simplified version of the master algorithm. It shows the functions that the algorithm must invoke for each FMU in the model. The figure reports only the execution of an initialized FMU that already successfully went through the FMU setup state. Once a FMU has been initialized, its execution reaches the *Step Completed* atomic state within the *State Initialized* sub-machine. The master algorithm may invoke the `fmi2Get` or the `fmi2Set` functions, respectively, reading or writing values of the FMU external variables. Otherwise, the algorithm may invoke the `fmi2DoStep` function by passing as a parameter the amount of time that must be simulated. Then, the machine moves to the *Step in Progress* state. The FMU simulates by executing its functionality: if the step is

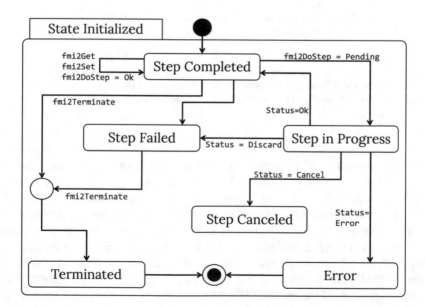

Fig. 2 Statechart representation of the coordinator algorithm for a FMU

not canceled or discarded, and no errors are caught during the FMU execution, the `fmi2DoStep` returns and the machine goes back to the *Step Completed* state, and the FMU advances its own local time according to the one previously passed as a parameter. These steps iterate until no `fmi2Terminate` function is invoked. A simulation tool may implement the simplest coordinator for FMI by iterating this process for each FMU, or it may implement some more complex mechanism, still adhering to the statechart in Fig. 2. Finally, the standard explicitly states that it is not legal to call a `fmi2Get` function after `fmi2Set` functions without calling the `fmi2DoStep` in between.

Listing 1 Sketch of the C implementation of a basic master algorithm compliant with the FMI standard. The algorithm executes a thousand iteration, each of those advances the local and global time of 10 time units

```c
int main(int ac, char * av[]){
  fmi2Component component_1 = load_fmu("./component_1.fmu");
  fmi2Component component_2 = load_fmu("./component_2.fmu");
  ...
  fmi2Status st;
  ...
  st = fmi2SetupExperiment(component_1);
  st = fmi2SetupExperiment(component_2);
  ...
  time = 0; step = 10;
  ...
  fmi2Integer in_1, in_2, out_1, out_2;
  // Simulation starts here.
  for(int i = 0; i < 1000; ++i) {
    st = fmi2GetInteger(component_1, 0, &out_1);
    st = fmi2GetInteger(component_2, 0, &out_2);
    in_1 = out_2; in_2 = out_1;
    st = fmi2SetInteger(component_1, 1, in_1);
    st = fmi2SetInteger(component_2, 1, in_2);
    st = fmi2DoStep(component_1, time, step);
    st = fmi2DoStep(component_2, time, step);
    time = time + step;
  }
}
```

Listing 1 shows a C implementation of a trivial master algorithm using the functions defined by the FMI standard for co-simulation. The procedure loads the FMUs instantiating two variable of type `fmi2Component` that will point to the FMU implementations (Lines 3–6). The status variable is declared (Line 8): every function defined in the standard returns a status. The master algorithm initializes the FMUs, the timing variables, and defines four integer variables (Lines 10–15). Then, a thousand simulation cycles are executed: the algorithm reads the output from the FMUs and assigns it to the input variables (Lines 18–20). Then, it sets the input variables of the FMUs (Lines 21–22). Finally, the algorithm executes the functionalities, advancing the global time of the FMUs and updates the global time (Lines 23–25).

2.3   Related Work

Some of the FMI standard weaknesses have been first identified in [5]. The main issues concern the managing of hybrid and discrete-event systems. The analysis highlights how FMI is more suited for physical, continuous-time (or discretized) systems, rather than discrete-event systems. Thus, it is tricky to use FMI when models require discrete events.

The semantic gap between continuous-time models and discrete-event models in FMI has been addressed [19] by proposing to use tokens synchronizing the FMUs in the model when discrete events happen. However, such a mechanism introduce many synchronization points in the execution, thus slowing down the simulation. This may be particularly inconvenient when simulating models coming from HDL descriptions, as we showed in [15]. In the same work we proposed an ad hoc synchronization methodology to reproduce the cycle-accurate behavior of HDL descriptions. It manages the synchronization locally to each FMU, while the data are exchanged by an additional FMU acting as a communication hub for the data in the system. The approach relies on automatic code generation to generate the FMUs implementing such mechanism. Automatic code generation of FMUs for co-simulation from HDL descriptions has been presented in [8]: it relies on a state-of-the-art abstraction technique [20] to translate HDL models into C descriptions. The generated descriptions are finally wrapped by an interface using the FMI co-simulation API.

While none of the approaches described above is proposing modifications to the standard, a number of papers do it. Cremona et al. [10] propose an additional mechanism to add to the FMI standard, aside from the model exchange and the co-simulation mechanisms. The novel mechanism is called *Hybrid Co-simulation*, and it is thought to manage hybrid models. The authors of [14] proposed some modifications to the API specified by the FMI standard for co-simulation. In particular, they proposed adding an *interrupt and preempt* mechanism to the fmi2DoStep. It allows the execution of an FMU to be interrupted when events must be managed.

To the best of our knowledge, **none of the previous work proposed to raise the abstraction level of FMUs to the** *transactional level*. This is due to the fact that the master algorithm must always know in advance the next step size for each FMU [4, 5]. This chapter shows how we overcame this limitation, enabling transactional level Cyber-Physical Virtual Platforms assembled relying on more abstract FMUs.

3   FMI Standard Advantages and Limitations

As a first contribution of this paper, we discuss the standard's features useful to create cyber-physical virtual platform. Then we will discuss some limitations that make integration of virtual platforms difficult. Our discussion will be from a "cyber" point of view, as we aim at highlighting the weaknesses of the standard when dealing with discrete-event and cycle-accurate components.

Indeed the standard allows to ease the integration of different tools. It simplifies the interfacing of heterogeneous description. It allows the designer to care only marginally about communication and synchronization between simulators. Furthermore, it is reasonable to assume that complex CPSs are designed by multiple teams of designers. For instance, a team might be in charge of the physical part, while the other designs the computational infrastructure. The FMI standard allows to easily integrate the models produced by different teams, to build a holistic simulation of the system.

However, as hinted in Sect. 2.3, the standard has been strongly oriented to continuous systems and dynamics. We can identify different drawbacks when modeling discrete components, and in particular when simulating digital components.

The set of *data types* provided by the standard is limited. When modeling digital HW it happens to use multi-valued logic values or signals that use an arbitrary number of bits. Meanwhile, FMI allows only integer, real, string, and Boolean. Thus, HDL data types must be mapped on the provided types. Different mappings have been already proposed in the past. Multi-valued logics and arbitrary long bit vectors have been mapped onto strings [7], and (more efficiently) abstracted to unsigned integer [8]. Still, none of the previous mapping is ideal even though they partially solve the problem.

The data types provided by FMI are even more insufficient when modeling digital HW models at higher levels of abstraction or when modeling SW. In such case, models may require aggregate data types, e.g., to represent sockets' payloads in transaction-level description, or classes of SW models. In this case, FMI does not provide any other solution than breaking down any aggregated type into its basic components.

The standard does not provide any mechanism to *specify the Model of Computation* employed by the FMU to implement the functionality. In the case of a digital HW description assignments are concurrent. However, simulators usually rely on sequential models of computation (e.g., data-flows). When aggregating digital HW components using FMI, complex synchronization structures must be built [15, 19] to guarantee the functional equivalence of the aggregated model of the system.

It is not possible to retrieve the *internal time of an FMU*. The master algorithm "imposes" to each FMU its internal timing. The main issue is related to the fmi2DoStep function behavior: it is called by the master algorithm and it carries on the simulation time while executing an FMU functionality. The execution of an FMU cannot be preempted by external events. Neither the FMU is allowed to simulate an amount of time different with respect to the one imposed by the master algorithm, since the FMU cannot communicate back to the master algorithm its effective internal timing. For this reason, the master algorithm must always be able to know exactly the length of the next time step of each FMU. This forces the master algorithm to call the fmi2DoStep function of an FMI using the shortest time step available or to perform multiple step revisions. Thus, this limitation leads to a higher number of synchronization points in the simulation and makes impossible to use advanced synchronization techniques, such as *temporal decoupling*. Thus, it is not well suited to manage discrete events that might be generated by system's components. In the case of HW description, this usually forces to simulate each FMU with a time granularity equal to the clock cycle [15].

4   Methodology

Figure 3 summarizes the proposed methodology. It starts from a set of HDL Intellectual Propertiess (IPs) models. An approach we precedently presented [8] (i.e., red box in Fig. 3) was simply translating the IPs into C models then wrapped into FMUs. Here we present a more advanced approach where HDL IPs undergo an abstraction and manipulation process (colored arrows in Fig. 3). The produced models rely on a transaction-level synchronization mechanism. Finally, these FMUs are inserted within the Cyber-Physical Virtual Platform, where they will be coordinated by a master algorithm that is aware of the shifting in synchronization and communication granularity achieved by applying the transformations.

4.1   FMUs Generation and Timing Backward Propagation

As identified in Sect. 3 FMUs cannot propagate their local time back to the coordinator. This is a major issue that must be tackled to achieve an efficient

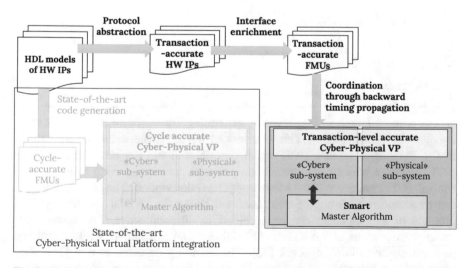

Fig. 3 Overview of the proposed approach, and comparison with the state-of-the-art methodology presented in [8]

discrete-event simulation. In fact, solving such issue will allow the master algorithm to decide the next simulation step length more efficiently. Furthermore, it will allow each FMU to simulate in a decoupled way, without defining the simulation step size. As such, when the Master Algorithm calls the fmi2DoStep, a FMU can simulate until it does not need to synchronize or communicate with other system components.

Listing 2 modelDescription.xml file of the component_1 with time port

```
1  ...
2  <ModelVariables>
3
4  <!-- Input Ports -->
5    <ScalarVariable name="in_1"
6                    causality="input" \
7                    valueReference="0">
8                    <Boolean start="false"/>
9    </ScalarVariable>
10
11   <ScalarVariable name="in_2"
12                   causality="input"
13                   valueReference="1">
14                   <Boolean start="false"/>
15   </ScalarVariable>
16
17 <!-- Output Ports -->
18   <ScalarVariable name="fmi2TLifaceTime"
19                   causality="output"
20                   valueReference="-1">
21                   <Integer start="0"/>
22   </ScalarVariable>
23
24   <ScalarVariable name="out_1"
25                   causality="output"
26                   valueReference="0">
27                   <Integer start="0"/>
28   </ScalarVariable>
29
30   <ScalarVariable name="out_2"
31                   causality="output"
32                   vr="1">
33                   <Integer start="0"/>
34   </ScalarVariable>
35
36 </ModelVariables>
37 ...
```

The proposed methodology starts by generating Transaction-Level models starting from HW descriptions. This is achieved by using the methodology defined in [3]. It takes as input a HDL model described at the Register Transfer Level (RTL) together with its communication protocol, and it generates a functionally equivalent Transactional Level Modeling (TLM) description. The HW descriptions can be provided by using the most common HDLs (i.e., VHDL or Verilog). The protocol of a component can be specified in different ways. The state-of-the-art implementations of the RTL-to-TLM abstraction methodology rely on ad hoc protocol specification languages [20]. The resulting description is a C++ class representing a Transaction-Level model of the original component. Each transaction of the system is executed by invoking its simulate function, and it emulates one transaction of the specified protocol. The internal time of the model is annotated as

`Integer` data type, which represents the number of clock cycles executed in the last transaction. The abstraction procedure computes the number of clock cycles for each transaction, and annotates it within the generated model.

The interface of the model is isolated in a structure embedded inside the C++ class. The structure contains a set of fields representing the original ports of the HW models. The data types of these fields are abstracted into C native data types. For instance, a 32-bit `logic_vector` data type is abstracted into `uint32_t` C data type. The methodology relies on automatic abstraction of HDL data type [20] to perform this transformation. Furthermore, the interface structure also contains the time annotation of the model.

Our methodology goes on by wrapping the generated C++ class within the FMI functions. Thus, it generates the set of `fmi2Set` and `fmi2Get` necessary to write and read, respectively, input and output variables from and to the components. It also generates the `fmi2DoStep` function that calls the generated `simulate` function emulating a component transaction. The `fmi2DoStep` function still accepts the step length to stay compliant with the standard. However, it ignores it as the actual internal time of the FMU is computed by the `simulate` function.

The methodology generates also the XML file for the FMU. The original ports of the HW model are mapped in the FMI data types: the `Boolean` and `Integer` FMI types are used to represent, respectively, single bit (or logic) and bit (or logic) vectors. The *value reference* is assigned to each port starting from 0 for each data type. Listing 2 depicts the definitions of the ports in the XML file for a component originally having two input and two output ports.

The methodology enriches the interface of the FMU with the internal time annotation of the transaction-level model that is exposed as a new `Integer` port of the FMUs (see Listing 2, Lines 18–22). The value reference -1 is reserved for the timing port. This assures that it can be uniquely identified once the FMU is loaded by a simulator. Furthermore, the timing port is called `fmi2TLifaceTime` in order to decrease the chances of name clashing with the other ports of the FMU. This last solution is helpful to increase also the readability of the produced FMUs.

4.2 A Better Coordinator for Discrete Systems

Listing 1 depicts a trivial Master Algorithm able to execute cycle-accurate FMUs. It must synchronize the components of the system at each clock cycle. Thus, the time step of each `fmi2doStep` is set to be equal to the clock period of the system being modeled. Such a solution is indeed precise; however, it uses an unnecessarily high number of synchronization points. The backward propagation of the FMUs internal time can be exploited to reduce the number of synchronization points.

Figure 4 shows the execution scheme of the proposed *Smart Master Algorithm*. Its core is the *FMU Coordinator*: it is in charge of storing the internal time values of the FMUs in the system, and it decides at each simulation step which components must be executed. Initially, the *Smart Master Algorithm* simulates all the FMUs,

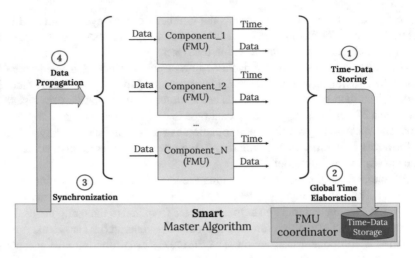

Fig. 4 Scheme of the *Smart Master Algorithm* with the *FMU Coordinator* of Transaction-Level FMUs

without defining a step size. All the FMUs return to the coordinator their internal time after their first execution. Then, the *Smart Master Algorithm* iterates the following steps (as in Fig. 4):

- ① *Time-Data Storing*: the internal time and the new data of each FMUs are retrieved from the master algorithm and passed to the *FMU Coordinator* that stores them.
- ② *Global Time Elaboration*: the *FMU Coordinator* elaborates the new Global Time of the simulation as the minimum value among all the internal times of the FMUs.
- ③ *Synchronization*: any FMU having the internal time equal to the Global Time is inserted into the list of *runnable FMUs*. Data read after the last execution of each runnable FMU, and previously stored by the coordinator, are shared with the system (i.e., the values become valid for the entire system).
- ④ *Data Propagation*: the *Smart Master Algorithm* propagates the data and simulates the FMUs present in the list of runnable FMUs.

Listing 3 Sketch of the C++ implementation of the *Smart Master Algorithm* exploiting backward timing propagation

```
1  ...
2  fmi2Status st;
3  unsigned int global_time=0;
4
5  fmi2Component components[num];
6  components[0]=load_fmu("./component_1.fmu");
7  components[1]=load_fmu("./component_2.fmu");
8
```

```
 9  unsigned int local_time_vector[num];
10
11  for(int i=0; i < num; i++) {
12    st=fmi2DoStep(components[i], global_time, 0);
13    st=fmi2GetInteger(component[i], -1, &local_time);
14    local_time_vector[i]=local_time;
15    retrieve_and_store_output(component[i]);
16  }
17
18  while(global_time < 1000) {
19    set< fmi2Component > runnable_FMUs;
20    global_time=find_minimum(local_time_vector[0]);
21
22    for(int i=0; i < num; i++) {
23      if(local_time_vector[i] == global_time)
24        runnable_FMUs.insert(components[i]);
25    }
26
27    propagate_data(runnable_FMUs);
28    set< fmi2Component >::iterator it;
29    for(it=runnable_FMUs.begin;
30        it != runnable_FMUs.end; it++)
31    {
32      fmi2Component * component=*it;
33      st=writeInputs(component);
34      st=fmi2DoStep(component, global_time, 0);
35      st=fmi2GetInteger(component[i], -1, &local_time);
36      local_time_vector[i]=local_time;
37      retrieve_and_store_output(component[i]);
38    }
39  }
40  ...
```

Listing 3 shows a sketch of the proposed Smart Master Algorithm. It reports only the most important parts of a possible C++ implementation of the coordination mechanism. Initially (Lines 2–9) it declares a status variable, an integer variable tracking the global time, and an array of components. The FMUs composing the system are stored in the array after being loaded. Furthermore, an array is declared to store the local times of each FMU. The same position in the two arrays refers always to the same FMU. Then, the coordinator initializes the simulation (Lines 11–16) by executing all the components once without advancing the global time. This step allows to generate the first set of events of the system, thus firing the event-based simulation mechanism and populating the set of runnable FMUs. For each execution, the local time is retrieved (Lines 13) and stored (Line 14). Then, all the output values written by the FMU are retrieved and stored (Line 15). Then, the system is simulated (Lines 18–39). At each simulation cycle a set containing the runnable FMUs is created empty and populated after the global time has been updated (Lines 19–25). Then, data previously produced by the runnable FMUs are propagated (Line 27). Finally, each runnable FMU is executed (Lines 28–37).

5   Methodology Application

We implemented the methodology by assembling a tool-chain performing the abstraction, manipulation, and translation steps. We relied on the API provided by the HIFSuite framework [2] to extend the automatic code generation presented in [8]. The automatic abstraction of HDL descriptions is performed by specifying the components' protocols to generate the corresponding transaction-level C++ descriptions as defined in [3]. The models produced by the abstraction are enriched with the timing backward propagation mechanism. Finally, a tool wraps the model within the FMI APIs for co-simulation. We applied the tool-chain to a set of benchmarks varying with respect to two dimensions: the protocol latency and the number of FMUs composing the system. We aim at estimating the scalability of the proposed approach with respect to these two dimensions. We implemented the same functionality within each HW component of the system, since this paper focuses on the interfaces of the components, rather than on their internal functionalities. The internal functionality is kept extremely simple in order to let the communication and synchronization overhead to be predominant in the simulation. Each component is simply counting the number of clock cycles until its pre-defined latency is reached. For each experiment, we have considered components with different latencies. In the experiments we refer to the *base latency* of an experiment as the minimum latency of the component in that experiment.

We generate two FMUs of different types for the same HW model: the cycle-accurate FMU and the transaction-level FMU with backward timing propagation. All the experiments have been performed on a 64-bit machine running Ubuntu Linux 16.04, equipped with 16 GB of memory and an Intel(R) Core(TM) i7-3770 CPU @ 3.40 GHz.

Table 1 reports the execution time by using the Trivial Master Algorithm, with different cycle-accurate FMUs and different numbers of iterations. The protocol latency dimension is not considered in this table because the Trivial Master Algorithm simulates only cycle-accurate FMUs. Using the Trivial Master Algorithm the protocol latency does not affect the coordination overhead in the simulation. The results show that moving in both the dimensions (number of FMUs or iterations) the execution time increases almost linearly.

Table 1 Execution time of FMUs simulation using trivial Master Algorithm, with different number of iterations

# iterations (clock cycles)	Execution of FMUs (s)				
	2	5	10	20	40
100 K	4.76	10.75	21.89	43.34	82.46
1 M	41.87	104.13	198.74	405.25	834.34
10 M	421.93	1021.64	2015.55	4129.17	8322.22
20 M	886.78	2062.32	4267.29	8219.65	16,466.54

Table 2 Execution time comparison of normal master algorithm and smart master algorithm with different protocol latencies

Base latency (clock cycles)	Execution of FMUs (s)									
	2		5		10		20		40	
	Trivial	Smart	Trivial	Smart	Trivial	Smart	Trivial	Smart	Trivial	Smart
20	421.93	115.54	1021.64	250.23	2015.55	465.48	4129.17	889.39	8322.22	1752.71
Speed-up	3.65×		4.08×		4.33×		4.64×		4.75×	
50	421.93	60.28	1021.64	135.65	2015.55	253.43	4129.17	481.36	8322.22	964.92
Speed-up	7.00×		7.53×		7.95×		8.58×		8.62×	
100	421.93	44.71	1021.64	95.57	2015.55	179.34	4129.17	344.25	8322.22	702.17
Speed-up	9.44×		10.69×		11.24×		11.99×		11.85×	

In all the scenarios, 10 million clock cycles of the system have been simulated

Table 2 compares the simulation speed achievable by using the Trivial and the Smart Master Algorithm. The performance obtained by using the Smart Master Algorithm depends on the protocol latency. On the contrary, the Trivial Master Algorithm performance is not influenced by such dimension. The Smart Master algorithm with the transaction-level FMUs achieves up to 11× speed-up when using the largest protocol latencies considered. Reducing the protocol latencies of the transaction-level FMUs, the Smart Master Algorithm is less beneficial because of the increasing number of synchronization points. Of course, when the protocol latency is equal to one clock cycle (e.g., when modeling combinatorial circuits) we have a degenerate case: the transaction-level and the cycle-accurate implementations will have the same amount of synchronization points. As such, only in that case, the Smarter Master Algorithm is slightly outperformed by the trivial one, due to the higher amount of computation required by the coordinator.

Figures 5 and 6 give a graphical representation of how the simulation overhead changes when changing the protocol base latencies and the number of FMUs, respectively. The vertical axes of both table report the simulation time, while the horizontal axes report the two considered dimensions. The trends in Fig. 5 show how performance improves by increasing the latency. This is because a longer latency allows for more temporal decoupling, thus less synchronization and communication overhead. Figure 6 shows that the simulation time increases linearly with the number of involved FMUs. Thus, it shows the minimal impact of the more sophisticated master algorithm proposed in this paper.

6 Recent Development and Discussion

In 2018 the FMI Steering Committee has announced a new version of the standard, called FMI 3.0. The committee also published the list of new standard intended

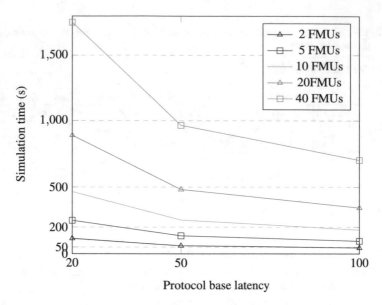

Fig. 5 Trend of the simulation overhead using the Smart Master Algorithm with respect to the protocol latency

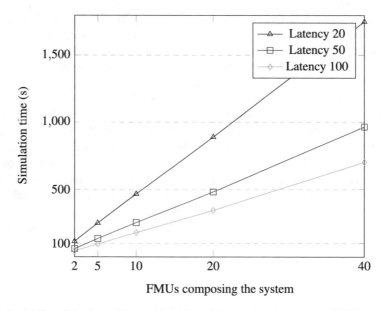

Fig. 6 Scalability of the Smart Master Algorithm with respect to the number of FMUs

additional features.[1] In our opinion, among the proposed features, the most interesting aiming at providing a better support for digital components are:

- new data types;
- structured ports and multi-dimensional variable;
- intermediate output values and support for hybrid co-simulation.

The Steering Committee also underlines that not all the mentioned features might be introduced in the final version of the new standard. Still, the entire project is now stored in a public repository,[2] it is thus possible to monitor the status of the development for the new standard. Currently, only new data types have been added to the features list. In details, it is now possible to specify integer values spanning from a single byte Integer (`fmi3Uint8`, `fmi3Int8`) to 64 bits Integer (`fmi3Uint64`, `fmi3Int64`).

Listing 4 Data types comparison between FMI Standard 2.0 and the new 3.0 versions

```
1  //FMI 2.0 DataTypes
2  typedef double          fmi2Real;
3  typedef int             fmi2Integer;
4  typedef int             fmi2Boolean;
5  typedef char            fmi2Char;
6  typedef const fmi2Char* fmi2String;
7  typedef char            fmi2Byte;
8
9  // FMI 3.0 Datatypes
10 typedef float           fmi3Float32;
11 typedef double          fmi3Float64;
12 typedef  int8_t         fmi3Int8;
13 typedef  uint8_t        fmi3UInt8;
14 typedef  int16_t        fmi3Int16;
15 typedef uint16_t        fmi3UInt16;
16 typedef  int32_t        fmi3Int32;
17 typedef uint32_t        fmi3UInt32;
18 typedef  int64_t        fmi3Int64;
19 typedef uint64_t        fmi3UInt64;
20 typedef int             fmi3Boolean;
21 typedef char            fmi3Char;
22 typedef const fmi3Char* fmi3String;
23 typedef char            fmi3Byte;
24 typedef const fmi3Byte* fmi3Binary;
```

Listing 4 shows a comparison between the data types provided by the standard 2.0 and the new standard 3.0. Only Lines 2–7 were present already in the previous standard. The new 3.0 standard extends the previously existing data types by adding the definitions reported in Lines 10–24. It is important to notice that some of the new definitions replace those of the former standard. For instance,

---

[1] https://fmi-standard.org/news/2018/05/30/fmi-3-0-alpha-feature-list.html.
[2] https://github.com/modelica/fmi-standard/.

`fmi2Integer` (Line 3) has been replaced with all the primitive C types that allow addressing specific amount of bytes (Lines 12–19). These new data types allow choosing between Signed or Unsigned and from a single byte to 64 bits Integer. *fmi2Real* (Line 2) has been splitted into `fmi3Float32` and `fmi3Float64` (Lines 10–11), where 32–64 represent the size of the data type (Floating-Point single precision or Double precision). Moreover, the introduction of these new data types implies consequentially the introduction of new methods for data-exchange (i.e., `fmi3GetUint8`, `fmi3SetUint8`, etc.). On the other hand, it is not clear if the new standard will provide compatibility with FMUs written by using the previous standards. One of the main purposes of the FMI standard is to support exchange of models among different teams, organizations, and tools. As such, it is our opinion that it will be necessary to provide interoperability between models produced by different organizations, or by the same in different times, and that may thus rely on different versions of the standard. Furthermore, we also think that keeping the possibility of using more generic types, such as generic integer or generic real, may help designers in the initial phases of the modeling process when some details may still be unknown. Moreover, it is our opinion that providing the possibility of using generic integer and real types may make the standard more attractive to users whose background is not in computer science. Another novel addition is the `fmi3Binary` data type. It is an opaque binary data type that may be useful to carry the information from complex sensors data to computational components, to model complex binary streams, or to model communication of closed-source components.

Structured ports and multi-dimensional variables are listed in the intended features list. Of course, supporting multi-dimensional variables will drastically move forward the standard toward the possibility of representing communication mechanisms typical of computing systems. In fact, it should provide the possibility of representing arrays, records, and other software typical data structures. The same will be true for hardware bus-based communication that can be represented by structured ports, similarly to what happens with the *payload* structures used in transactional models. For instance, structured ports will allow to simplify the approach presented in this chapter by using a single port representing the entire payload of the component modeled in a transactional FMU. However, at the current state of the work, the development of structured ports and multi-dimensional variables has been only announced, and any further detail has not been presented yet.

The possibility of accessing internal data in the middle of the `doStep` function is an extremely promising feature to enable hybrid co-simulation. In fact, accessing internal events of module being simulated enables the possibility of modeling mechanisms similar to interrupt that are crucial to model reactive systems. Furthermore, such a feature will enable a better managing of events and time. Since important events may be visible to the master algorithm even before the termination of a FMU execution, the master algorithm can speculate by increasing the length of the FMU execution. This feature will also make obsolete the solution proposed in this chapter: here, an FMU simulates until the first interesting event, and then the master algorithm must retrieve the internal FMU time by the additional port proposed in

Sect. 4. With the new feature, the master algorithm may impose a longer FMU execution while monitoring eventual internal events of interest, thus decreasing the number of required synchronization and communication points. Then, better mechanisms of handling the co-existence of FMUs governed by different models of computation, as well as mixed continuous- and discrete-time dynamics must be incorporated. However, even though some extensions have been proposed in the literature [10, 14], and at its current state, the implementation does not clarify if such extensions will be integrated into the new standard. Integrating such features will be crucial to support digital models within FMI-based simulation frameworks. Otherwise, users will continue to be forced performing sophisticated manipulations to models, such as those proposed in this chapter.

7 Concluding Remarks

This chapter discussed the current version of the FMI standard, and proposed a methodology to extend it to simulate FMUs representing digital components at transaction-level. The approach adds some information to the FMUs interface. Then, it adopts an ad hoc master algorithm that is still conformed to the standard.

The experimental results showed the positive impact of the methodology. However, the approach requires design effort to explicitly force the standard to accept the transaction-level FMUs we defined. The analysis of the current effort to extend the standard shows both the importance of the proposed extensions and the necessity to better target such extensions to support discrete-event models. Meanwhile, the proposed methodology will allow to cover where the current standard is still lacking.

References

1. Blochwitz, T., et al. (2012). Functional Mockup Interface 2.0: The standard for tool independent exchange of simulation models. In *Proceedings of MODELICA Conference 2012* (pp. 173–184).
2. Bombieri, N., Di Guglielmo, G., Ferrari, M., Fummi, F., Pravadelli, G., Stefanni, F., et al. (2010). HIFSuite: Tools for HDL code conversion and manipulation. *EURASIP Journal on Embedded Systems, 2010*(1), 1–20.
3. Bombieri, N., Fummi, F., & Pravadelli, G. (2011). Automatic abstraction of RTL IPs into equivalent TLM descriptions. *IEEE Transactions on Computers, 60*(12), 1730–1743.
4. Broman, D., Brooks, C., Greenberg, L., Lee, E.A., Masin, M., Tripakis, S., et al. (2013). Determinate composition of FMUs for co-simulation. In *Proceedings of the Eleventh ACM International Conference on Embedded Software* (p. 2).
5. Broman, D., Greenberg, L., Lee, E. A., Masin, M., Tripakis, S., & Wetter, M. (2015). Requirements for Hybrid Cosimulation Standards. In *Proceedings of the 18th International Conference on Hybrid Systems: Computation and Control* (pp. 179–188). New York: ACM.
6. Cai, L., & Gajski, D. (2003). Transaction level modeling: An overview. In *Proceedings of the 1st IEEE/ACM/IFIP CODES-ISSS* (pp. 19–24). New York: ACM.

7. Centomo, S., Deantoni, J., & De Simone, R. (2016). Using SystemC cyber models in an FMI co-simulation environment: Results and proposed FMI enhancements. In *Proceedings of Euromicro Conference on Digital System Design (DSD)* (pp. 318–325). Piscataway: IEEE.
8. Centomo, S., Lora, M., Portaluri, A., Stefanni, F., & Fummi, F. (2017). Automatic generation of cycle-accurate Simulink blocks from HDL IPs. In *Proceedings of ECSI/IEEE Forum on Specification & Design Languages 2017 (FDL 17)* (pp. 1–8).
9. Centomo, S., Lora, M., & Fummi, F. (2018). Transaction-level functional mockup units for cyber-physical virtual platforms. In *2018 Forum on Specification & Design Languages (FDL)* (pp. 1–8). Piscataway: IEEE.
10. Cremona, F., Lohstroh, M., Broman, D., Lee, E. A., Masin, M., & Tripakis, S. (2017). Hybrid co-simulation: It's about time. *Software & Systems Modeling.* https://doi.org/10.1007/s10270-017-0633-6.
11. Fummi, F., Lora, M., Stefanni, F., Trachanis, D., Vanhese, J., & Vinco, S. (2014). Moving from co-simulation to simulation for effective smart systems design. In *Proceedings of the conference on Design, Automation & Test in Europe* (p. 286). Leuven: European Design and Automation Association.
12. Golomb, S. W. (1971). Mathematical models: Uses and limitations. *IEEE Transactions on Reliability, 20*(3), 130–131.
13. Lee, E. A.: Fundamental limits of cyber-physical systems modeling. *ACM Transactions on Cyber-Physical Systems, 1*(1), 3 (2017).
14. Liboni, G., Deantoni, J., Portaluri, A., Quaglia, D., & De Simone, R. (2018). Beyond time-triggered co-simulation of cyber-physical systems for performance and accuracy improvements. In *Proceedings of workshop on rapid simulation and performance evaluation: Methods and tools.*
15. Lora, M., Centomo, S., Quaglia, D., & Fummi, F. (2018). Automatic integration of cycle-accurate descriptions with continuous-time models for cyber-physical virtual platforms. In *Proceedings of ACM/IEEE Design Automation & Testing in Europe 2018* (pp. 1–6).
16. Lora, M., Vinco, S., & Fummi, F. (2019). Translation, abstraction and integration for effective smart system design. *IEEE Transactions on Computers.*
17. MODELISAR Consortium, Modelica Association, et al.: *Functional Mock-up Interface for Model Exchange and Co-Simulation – Version 2.0.* Available from https://www.fmi-standard.org.
18. Rajkumar, R. R., Lee, I., Sha, L., & Stankovic, J. (2010). Cyber-physical systems: The next computing revolution. In *Proceedings of the 47th Design Automation Conference* (pp. 731–736). New York: ACM.
19. Tripakis, S. (2015). Bridging the semantic gap between heterogeneous modeling formalisms and FMI. In *Proceedings of International Conference on Embedded Computer Systems: Architectures, Modeling, and Simulation (SAMOS)* (pp. 60–69). Piscataway: IEEE.
20. Vinco, S., Guarnieri, V., & Fummi, F. (2016). Code manipulation for virtual platform integration. *IEEE Transactions on Computers, 65*(9), 2694–2708.

Safe Interoperability for Web of Things Devices and Systems

Ege Korkan, Sebastian Kaebisch, Matthias Kovatsch, and Sebastian Steinhorst

1 Introduction

The Internet of Things (IoT) brings connectivity to electronic devices and allows them to connect with each other. Due to the large variety of IoT devices and application scenarios, they all bring their own properties such as different processing speed or range of connectivity, desired run-time or energy consumption, safety features, etc. This creates a fragmentation in IoT, with different standards to interact with the devices and to represent them, each optimized for a specific application area or device type. Consequently, such fragmentation hampers composing applications beyond the functionality of the individual devices.

In the electronic design community, languages such as SystemVerilog have proven to be an effective standardized representation for the entire development cycle, from design to verification and for a very wide range of application areas. However, in the IoT domain, companies introduce siloed IoT platforms that come with proprietary standards even within similar application domains.

Consequently, there is a necessity that an IoT device can be represented with a description of capabilities, which can be understood and interpreted by other devices and standards. Here, a common ground can be created by enabling to describe an

E. Korkan (✉)
Technical University of Munich, Munich, Germany
e-mail: ege.korkan@tum.de

S. Kaebisch · M. Kovatsch
Siemens AG, Munich, Germany
e-mail: sebastian.kaebisch@siemens.com; matthias.kovatsch@siemens.com

S. Steinhorst
Technical University of Munich, München, Bayern, Germany
e-mail: sebastian.steinhorst@tum.de

© Springer Nature Switzerland AG 2020
T. J. Kazmierski et al. (eds.), *Languages, Design Methods, and Tools for Electronic System Design*, Lecture Notes in Electrical Engineering 611,
https://doi.org/10.1007/978-3-030-31585-6_3

47

interface to different standards in a well-defined representation. For this purpose, the Thing Description (TD) [1] was introduced recently as an open description format for devices with connectivity of any kind that is human-readable and machine-understandable. The TD is not a standard to replace other IoT standards, but it enables to describe them through syntactic and semantic information.

Consider a temperature sensor used with a cloud IoT platform and a local ventilator. Between them, TDs enable to create a temperature-controlled ventilation system directly composed of the capabilities of these two physical devices. The advantage of such interoperability for machine-to-machine communication is to enable system functionality without prior knowledge about the interfaces between the devices.

Such a sensor's functional capability, data structure, and access points will be referenced in the TD of the sensor. Hence, the ventilator will be able to access the sensor data due to the provided access points and will be able to understand the data due to the data structure described in the TD.

The previous ventilation system example is abstracted in Fig. 1. This system has three IoT devices, each possessing a TD. Within the system, each IoT device, to which we will in the following refer to as a Thing,[1] can read the TD of another Thing and interpret it to understand the information such as the Thing's interactions, supported protocols, data structure, how to access the data, etc., as described in the column on the right of Fig. 1 (TD Contents). During the course of this paper, an exposer Thing accepts requests provided in its TD, whereas the consumer Thing reads a TD and interacts with the exposer Thing.

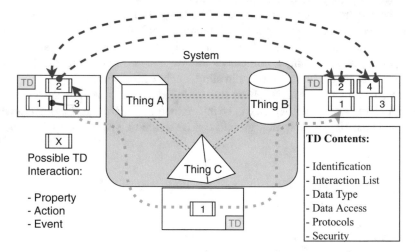

Fig. 1 An abstracted view of an IoT System with three IoT Devices each with an associated Thing Description (TD). The arrows demonstrate composition of greater functionality than the devices themselves, necessitating sequential behavior between devices

---

[1]When the word Thing is used with a capital letter, a Thing means an object, either virtual or physical, that can be communicated with.

An interaction is the description of a specific capability of the Thing, representing the data structure, access protocol, and access link. For example, reading the temperature value is such an interaction with the Thing. Similarly, rotating the fan is also an interaction that acts on the physical world. In a TD, one would find a list of interactions and how to access them. Interactions are illustrated by numbered boxes in Fig. 1 and they will be explained in Sect. 2 in more detail.

In Fig. 1, Thing A has three interactions and all these interactions can be used by Thing B and C to interact with Thing A. Referring to the temperature-controlled ventilation system example, interaction 1 of Thing A can be reading the temperature value and the interaction 4 of Thing B can be rotating the fan.

Problem Statement With the current TD standard, it is possible to build the system described in Fig. 1. However, the behavior represented by arrows has to be programmed manually, which results in an implicit description of the device or system.

An interaction can change the state of the Thing, making it accept only certain interactions (state transitions). For example, the red (continuous) arrow is a sequence describing such state transitions of Thing A. This can be requirements of sequential behavior, such as initializing the motor driver of the ventilator before setting a rotation speed. In order to execute this sequence of interactions, since such a sequence is not described in the TD, the person who implements the compositional system needs to have access to an operation manual of Thing A. This manual should describe the internal workings of the Thing (e.g., with a state machine) and give meaning to the causality between interactions.

Similarly, the green (dotted) and blue (dashed) arrows in Fig. 1 illustrate sequential behavior between multiple Things and are not expressed anywhere, thus need to be implemented manually. For example, we would like to express that the green (dotted) arrow represents the aforementioned temperature control functionality in the correct order and with a causal relation: reading a temperature value and then rotating the ventilator. This shows that executing multiple interactions can provide another meaning that is not previously given in a single interaction. To solve this problem, a new interaction can be implemented that provides the same meaning of executing multiple interactions. This is possible during the development phase of Things, but for non-reprogrammable, legacy devices there is no such option.

Contributions In order to avoid that each interaction is executable at any given time or multiple interactions can be executed in any given order, in this paper, we propose the specification of sequential behavior within TDs. The ability to represent valid sequences of interactions, which we call **paths**, in the TD of a device enables the designer of this device to restrict interactions and hence simplify the interaction of other devices with this device. Without such paths, arbitrary sequences of interactions could be triggered, which would either require knowledge about the inner workings of the device or create an unsafe and erroneous behavior.

Consequently, in the context of TDs introduced in Sect. 2, this paper has the following contributions:

- We extend our initial path vocabulary[2] contribution of [2] in Sect. 3.1 that uses the JSON Pointers instead of `hrefs`. This enables stronger semantics for describing how to interact with Things.
- We show that a system can be composed through sequential interactions of multiple Things by using the same **path** logic, presented in Sect. 3.2.
- We demonstrate a case study with sequential behavior in an industrial automation system composed of an industrial fan, a temperature sensor, and a system controller in Sect. 4.

Related work is discussed in Sect. 5, a discussion on an application of our methodology is provided in Sect. 6, and Sect. 7 concludes.

As the applications of TDs diversified, two new observations motivated us to improve our path vocabulary:

- Some Web of Things (WoT) devices use the same Uniform Resource Identifier (URI) as `hrefs` for multiple interactions where the interactions are differentiated via the method they require. For example, the Philips HUE [3] lights specify that sending an HTTP GET request to /api/<username>/lights would return all the light states and information, like a Property Interaction of a TD. However, sending an HTTP POST request to the same URI would start a search for new lights.
- In some cases, one interaction can have multiple forms that serve different purposes, such as one for observing possible updates of a sensor's measurement and one to get the current value. These forms can use the same `href` value or even use different protocols.

These new discoveries motivated us to abstract how the path vocabulary is serialized and not use URIs of `hrefs` in the path serialization. We have opted for the JSON Pointers standard as specified in [4] which is still in the URI format. JSON Pointers point to a specific place in a JSON document and in our contribution we use them to point to a specific `form` in the TD document.

2 Thing Description

The TD approach has been introduced in September 2017 (First public draft) by the WoT Working Group of World Wide Web Consortium (W3C). This section will explain the TD approach, but most importantly, its shortcomings and why our contribution is necessary to enable TDs to describe more complex cyber-physical

---

[2]The term vocabulary is used here since the TD standard [1] refers to actions, properties, etc. as a vocabulary.

systems. In the following, we will mainly focus on the relevant details of TDs for the context of our contribution, the proposed path vocabulary.

The path vocabulary that will be introduced in Sect. 3 describes a series of interactions. Further information on the characteristics of interactions is thus required before introducing this vocabulary. In this section, we will define interactions in order to argument the need for describing sequential behavior.

An interaction I can represent two types of messaging patterns: request–response (Definition 1) and publish–subscribe (Definition 2).

Definition 1 (Request–Response) For a request $p \in$ client and a $q \in$ server, the pair is defined as follows:

$$p \Rightarrow q \tag{1}$$

Definition 2 (Publish–Subscribe) Notifying an event only in matching subscription intervals is defined by Baldoni et al. [5] as follows:

$$\forall e \in \text{nfy}(x) \in h_i \Rightarrow \text{nfy}(x) \in S_i(C) \text{ s.t. } C(x) = \top \tag{2}$$

with

- e, the event the subscriber subscribed to;
- x, the information generated from the process;
- nfy, the notification of the information;
- h, a local computation that generated x;
- S, the interval between subscription and unsubscription;
- C, the subscription request by the subscriber;
- $\top$, the pattern of the event to subscribe to at the server side.

These formal definitions for interactions are mentioned in the TD standard [1] in three groups:

- Properties: A value provided by the Thing, such as sensor data, or values provided to the Thing, such as a desired temperature. This matches the request–response pattern.
- Actions: Requesting the Thing to do something that interacts with the physical world or with other Things that also takes some time, such as turning on a fan or LED. This matches the request–response pattern.
- Events: A message triggered due to a change in the Thing and sent to the consumer Things that have subscribed to it, such as an overflow alarm. This matches the publish–subscribe pattern.

In order to illustrate the different types of interactions in a practical example, we are showing a simplified TD of a ventilator in Listing 1. This ventilation Thing, as described by its TD, can rotate the motor of the ventilator at a given speed provided by the consumer Thing. It also has safety features such as requiring initialization by the consumer Thing. In addition, in case of an overheating of the motor, it can notify the consuming Things who are subscribed to this notification.

Other than interactions, the TD provides identification information. In the order of appearance in Listing 1, the `title` provides a human-readable reference (identification) for this Thing, whereas `id` provides a unique identification for the Thing that stays unchanged through different networks or IP addresses. Similarly, the `base` (line 3) describes the protocol and the URI needed to communicate with this Thing.

By using the default protocol bindings described in [6], one can interact with the previously introduced ventilator in the following sequence:

- Read or write the rotation speed of the ventilator by reading/writing the `rotation` property (lines 6–10). Here, it is specified that the data structure should be an `integer`.
- Rotate the ventilator by invoking the `rotate` action (lines 13–15). This action can be invoked without sending any specific data and the response will not contain an `integer` as in the previous property.
- Initialize the motor driver by invoking the `initialize` action (lines 16–19). Here, it is specified that the data structure of the response should be a `string`.
- Subscribe to the `overheating` event (lines 22–25) and get notified if the motor heats up too much. The structure of the data received will be a `string` data structure.

This ventilation Thing represents a sequential behavior that is not explicitly described. If one reads and learns the internal workings of the Thing, it is specified that in order to rotate the motor, one needs to invoke the `initialize` action (lines 16–19). This problem is commonly encountered in cyber-physical systems and is illustrated in an abstracted fashion in Fig. 2. Generally, a consumer Thing reads a TD, understands what can be done with the associated Thing, sends a chosen request to execute the interaction, and waits for the response from the Thing. The orange (dashed) arrow `Choose Interaction` is thus handled implicitly by the Thing Y (consumer) and there is no vocabulary provided by Thing X that tells the consumer to execute interactions in a specific order. Without the contribution of this

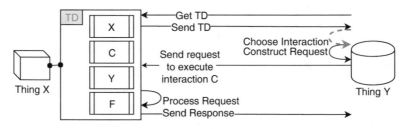

Fig. 2 Request–Response sequence abstraction that can be used for interacting with a Thing. The orange (dashed) arrow demonstrates the missing part of the TDs, which is the problem addressed in this paper

paper, Thing Y's developer had to know the internal workings of Thing X. With our contribution, presented in the following section, this becomes a more systematic and guided process.

```
 1 {
 2   "title": "MyVentilator",
 3   "@context": "https://www.w3.org/2019/wot/td/v1",
 4   "id": "urn:dev:ops:32473-ventilator-1234",
 5   "securityDefinitions": {
 6     "basic_sc": {"scheme": "basic", "in":"header"}
 7   },
 8   "security": ["basic_sc"],
 9   "base":"coaps://vent.example.com:5683",
10   "properties":{
11     "rotation":{
12       "type": "integer",
13       "forms":[{"href": "/rotation"}]
14     }
15   },
16   "actions":{
17     "rotate":{
18       "forms":[{"href": "/rotate"}]
19     },
20     "initialize":{
21       "output":{"type": "string"},
22       "forms":[{"href": "/init"}]
23     }
24   },
25   "events":{
26     "overheating":{
27       "data":{"type":"string"},
28       "forms":[{"href": "/oh"}]
29     }
30   }
31 }
```

Listing 1 Simple Thing Description of a ventilator that exposes the rotation speed, motor initialization, and rotating actions and an overheat alarm that can be obtained from coaps://vent.example.com:5683/td

3 Describing Sequential Behavior

The contribution of this paper is the new path vocabulary that allows to describe sequential behavior. We start this section by listing some requirements of such a vocabulary in the context of TDs. The following subsections start by introducing the vocabulary for single devices and then extend it for systems composed from devices.

Many models for system representation are measured by their expressiveness. In the field of automata theory, there are different levels of expressiveness, from finite automata to Turing Machines.

For cheap and not powerful IoT devices, exhaustive modeling of the inner workings is too tedious. On the other hand, a behavior described in a TD needs to be parsed and understood by such resource-constrained devices. Hence, even if the device providing this representation has enough resources to provide it, the description will not be usable by other IoT devices that are resource-constrained. Furthermore, obliging interacting devices to understand such behavior is contradictory to the design philosophy that internet and web technology enabled in the last decades, which is also applied for IoT.

Often, web pages, services, or Application Programming Interfaces (APIs) are self-descriptive and the user does not need to understand the complete system to start using them. For example, in a simple web page, the user can simply understand the link that he/she is interested in and not look at the rest (e.g., a site-map), i.e., not understand the complete state machine to execute one interaction. Inspired by the success of this logic, it is primordial to follow the same logic for IoT systems and hence for TD, in order to enable easy adoptability and usability.

3.1 Describing Sequential Behavior in a Single Thing

The path vocabulary is based on describing sequential behavior for a single Thing. For this reason, we will formally define the path vocabulary in this section. The formal definition will be then embedded into the TD format and later on used in a system. In order to illustrate the problem and guide this paper, we will be using a state machine of a legacy motor driver of a ventilator, as shown in Fig. 3.

This device cannot be reprogrammed,[3] but requires strict sequential behavior in order to operate safely. A sequence of interactions is needed to make it ready for accepting speed commands or to bring it back to a safe stop.

We can see that the `initialize` action needs to be invoked to initialize the motor. This sets the rotation per minute (rpm) of the motor to 0. However, as a safety feature, the `rotate` action must be explicitly invoked before setting the rotation speed with the `rotation` property. At this point, we can write to the speed value and rotate the motor in a direction. For example, to rotate the motor at 1300 rpm, the following specific order of interactions is needed:

1. Initialize
2. Rotate
3. Write (1300 rpm as value)

A consumer Thing that will interact with this motor driver and that does not know this sequential behavior cannot control the machine the way it is designed.

---

[3]TDs allow precise description of the capabilities of a device even if the device cannot provide its own TD. In this case, we can use a gateway that stores and provides the TD.

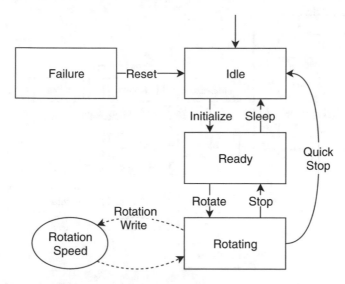

Fig. 3 State machine representation of a legacy motor driver. In order to enable setting the rotation speed to the desired value, Initialize, Rotate interactions have to be executed in this order

Furthermore, if the consumer Thing has access to this specific state machine in a machine-readable format (such as SCXML [7]), understanding the entire state machine for every application should not be necessary. For example, if the motor driver, i.e., the exposer Thing, chooses to expose only a safe stop sequence, the entire state machine that also describes the sequence to rotate the motor would contain unnecessary information.

By contrast, in our path vocabulary, we describe the behavior we want to expose with simple sequential interactions with interaction data that already exist in the TD. The aforementioned path of interactions, named `RotateMotor`, is shown in Fig. 4 along with the state machine from Fig. 3 that was used to generate the paths. We have given other valid path examples from the state machine for illustration.

In order to properly define the path vocabulary we need to introduce four definitions this vocabulary is composed of: path, name, @type, and paths.

Definition 3 (Path) From an ordered sequence of interactions I of sequence length l with $1 \leq i \leq l$, a path π with name t is defined as:

$$\pi_t = I_1, \ldots, I_i, \ldots, I_l \tag{3}$$

Definition 4 (Name) The name of the path is used within the TD to reference the JSON [8] object that contains the path information. Within the TD, the name allows the path to be referenced in the following fashion:

$$\pi_t = \text{derivePath}(t) \tag{4}$$

with derivePath being a function that finds the path t by parsing the TD.

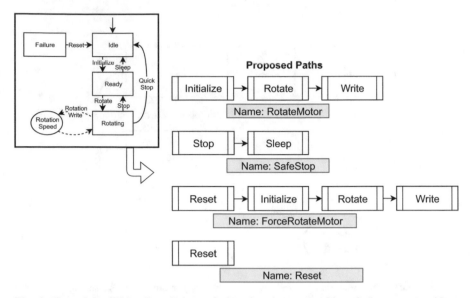

Fig. 4 Illustration of Thing Description paths based on the state machine of a legacy motor driver for an industrial fan. The paths are composed of interactions that execute state transitions. Note that even if the path just contains a single interaction, it is still a valid representation

Definition 5 (@type) The `@type` optionally allows to annotating semantics with the path. It uses the JSON-LD [9] format to reference to another resource on the Web that gives a meaning to the path, making it machine-readable. In a TD, this semantic annotation is given in a compacted form. The value written in `@type` will be combined with a URI in the `@context` field of the TD, exactly the same way as it is combined in the TD standard [1]. Currently used semantic annotations can be found in the iot.schema.org library[4] and used for linking the data.

Definition 6 (Paths) The set of paths offered by the Thing is denoted by Π and defined as follows:

$$\Pi = \bigcup \pi_k \mid \pi_k \in \text{TD} \tag{5}$$

These formal definitions translate to a path description in a TD as shown in Listing 2. Paths are an extension of the TD in Listing 1, with ... symbolizing the interactions of this TD. This specific TD offers only two paths: `rotateMotor` to rotate the motor from an initial state by executing `initialize`, `rotate`, and `rotation`, as well as `safeStop` that brings the motor to the initial state by executing `stop` and `sleep`, in these respective orders.

---

[4]http://iot.schema.org/.

The paths in the TD are serialized using the JSON Pointers standard [4]. JSON Pointers are URIs, so they can be easily parsed by Things. In Listing 2, the paths have relative JSON Pointers, where the # sign points to the root of the TD document. After this sign, the path points hierarchically to the form that is a member of the path.

Dealing with Legacy Devices TDs are envisioned for any device that needs to be connected to an IoT system. As we have mentioned before, the motor driver of the ventilator is a legacy device. During the course of this paper, we have used *modern* protocols such as CoAP [10] in the TD listings. However, the advantage of TDs is the capability to describe also older protocols such as Modbus [11], widely used in industrial automation. Such devices might be also non-reprogrammable, which means that they cannot provide a TD themselves. In this case, the TD of such a device has to be retrieved from a database. Thus, the TD of the ventilator has been retrieved from a local database and used by a gateway.

The use of a gateway is necessary to provide access to the functionalities of the legacy device to devices that do not have direct access to the legacy device, such as not supporting the protocol of the legacy device or not having a physical connection. Such a configuration is illustrated in Fig. 5 with Thing C as the device that does not have direct access to Thing A, the legacy device.

The gateway can then proceed on making the paths of the legacy device simple to use for consumer Things, such as Thing C. In the context of IoT, path descriptions should not be imposed to consumer Things that are not part of the system.

We are expecting to see our path vocabulary to be used inside the system and not in the TD of a device such as a gateway. Hence, the TD of the gateway should present simple interactions that should be executable without any causality. In Fig. 5, the path RotateMotor becomes an interaction with the same name that will be executed as a normal TD interaction by Thing C.

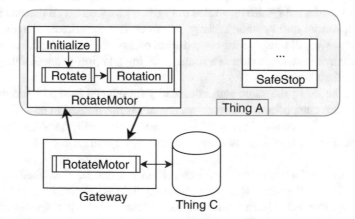

Fig. 5 Using a gateway brings IoT connectivity to a legacy motor driver (Thing A). The gateway can execute a path offered by this device and offer a simple Thing Description action to be executed by Things that do not have physical access to Thing A, such as Thing C

```
 1 {
 2   "name": "MyVentilator",
 3   ...
 4   "paths":{
 5     "rotateMotor":{
 6       "@type":"iot:rotate",
 7       "path":[
 8         "#/actions/initialize/forms/0",
 9         "#/actions/rotate/forms/0",
10         "#/properties/rotation/forms/0"
11       ]
12     },
13     "safeStop":{
14       "@type":"iot:stop",
15       "path":[
16         "#/actions/stop/forms/0",
17         "#/actions/sleep/forms/0"
18       ]
19     }
20   }
21 }
```

Listing 2 Thing Description of the motor driver with the paths that represent the interaction sequences

3.2 Composing a System

In the context of IoT, we are considering resource-constrained devices that are not able to offer a lot of functionality on their own. This is why composing a system by bringing multiple devices together to orchestrate more functionalities is highly relevant. Consider the system illustrated in Fig. 6, with a Thing B that can measure room temperature and another Thing A, which is a ventilator, to reduce room temperature. We will illustrate the composition of a system by using the two devices that can control the temperature of a room, bringing additional functionality just by combining their abilities.

We will be using the same path vocabulary introduced in the previous section for this system composition. The path vocabulary is not limited to describe a single Thing, but can be used for a system of Things and the causality between interactions of multiple Things. By using the same vocabulary, we will enable a scalable design approach.

The aforementioned temperature control system can be described by simply using the JSON Pointer URIs from different TDs to describe a system level functionality in a path. Such a path can be executed through a system controller or a Thing of the system. Figure 6 illustrates this system with a system controller where the gateway device takes the responsibility of describing the system behavior and executing system level functionalities.

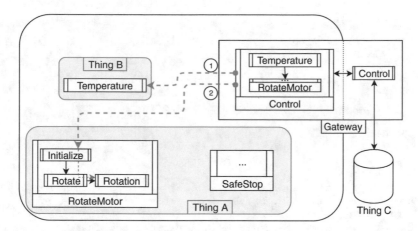

Fig. 6 A gateway can compose a system with the use of the path vocabulary. Here, the system is a temperature control system with a temperature sensor and an industrial ventilator. Things, such as Thing C, that do not have physical access to the system components can execute simple Thing Description interactions to interact with the system through the gateway

The dashed orange arrows in Fig. 6 demonstrate a path executed by the system controller. The system controller is thus able to execute paths or interactions of other devices due to its system controller TD.

Since a path can be also referenced, like an interaction form, with a JSON Pointer, a path and an interaction can be mixed into another path. This is illustrated in Fig. 6 by the `control` path that has the `temperature` interaction and the `rotateMotor` path combined. This means that our path vocabulary can scale well and create a compositional design flow for IoT systems. Listing 3 shows the TD of the gateway illustrated in Fig. 6. The path called `control` can either be offered as an interaction to the consumers of the gateway or directly used, just as the gateway is using the path of the ventilator. As a result, based on thoroughly tested simple interactions and paths, more complex behavior can be described and offered to higher level system controllers.

Note that the URIs have to be absolute URIs in a system controller, since relative URIs lose their uniqueness outside the TD.[5]

---

[5] A path URI in a TD such as `#/actions/initialize/forms/0` can be combined with the URI of the TD to create a URI that is valid also outside a TD. In this case, it would be coaps://vent.example.com:5683/td#/actions/initialize/forms/0.

```
1 {
2    "id": "urn:dev:ops:32473-controller-1234",
3    "title": "SystemController",
4    "@context": [
5       "https://www.w3.org/2019/wot/td/v1",
6       {
7          "iot": "http://iot.schema.org/iot"
8       }
9    "paths":{
10      "control":{
11         "@type":"iot:temperatureControl",
12         "path": [
13   "http://fdlSensor.com:5683/td#/properties/temperature/forms/0",
14   "coaps://vent.example.com:5683/td#/actions/initialize/forms/0",
15   "coaps://vent.example.com:5683/td#/actions/rotate/forms/0",
16   "coaps://vent.example.com:5683/td#/properties/rotation/forms/0"
17      ]
18      }
19   }
20 }
```

Listing 3 Thing Description of a system controller/gateway of the temperature control system with a path composed of URIs of interactions of system components

3.3 Worldwide Scalability

As seen in Listing 3 the path URIs can contain domain names that are globally available. These domain names resolve to a particular IP address of a device belonging to the system. However, this device can belong to any network in the world since it is an Internet connected device.

This illustrates that the TD can be used to represent any device in the world; thus, paths can describe behavior of a system composed of devices anywhere in the world. In Fig. 7, we illustrate such a system where a central controller can interact with single Things like Thing C (bottom right), systems like in Fig. 6 (top right), or with virtual Things in the cloud like Thing X (top left). In this scenario, the central component is able to compose a water level monitoring service that gets weather predictions from a virtual Thing in the cloud in another location, can combine with controlled temperature from a third location, and finally control the water level by pumping water in a fourth and final location.

4 Case Study: Testing with Path Semantics

Ideally, a TD describes what a Thing can do, but it is up to the developer of the Thing to properly implement the capabilities. It is even more difficult to implement everything correctly when designing and implementing a system because of the

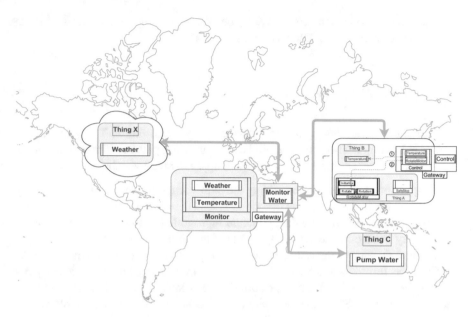

Fig. 7 Another IoT system using other systems, devices, and cloud to compose itself. The ubiquity of Internet and World Wide Web allows Thing Descriptions and the path annotations to be scalable on a worldwide scale

interlinked behavior of devices that compose the system. During both development processes for single Things as well as for systems of Things, testing becomes helpful to detect any errors in the implementation. However, manual testing is a tedious process and for this reason, automatic testing methods are widely used in many application domains.

In a case study, we will show how to apply TDs with the new path vocabulary to facilitate automated testing. In order to show the advantages of our contribution, we will compare the test coverage of our new path-enabled approach to the state-of-the-art testing without paths through an example. Similar to the previous section, we will first present this for a single Thing and then for a system. In the end, an algorithm that is applicable to test both single Things and systems will be shown.

TDs, with or without the path vocabulary, describe exposer Things that the consumer Things will interact with. Since a TD is human-readable, it can be used for specifying a Thing to develop (product), read by the developers who are not familiar with the internal workings of the device during implementation and more importantly, since it is machine-understandable, it can be used for automatic testing to generate test scenarios.

In the following, for automatic testing, we will use the black-box testing approach. In black-box testing, inputs are given to a device under test and the outputs are observed. This type of interaction is equal to a consumer Thing interacting with an exposer Thing. Since the consumer interacts with the exposer based on the information obtained from its TD, black-box testing of an exposer Thing implementation can be automatized by using its TD.

4.1 Single Thing Testing

We will demonstrate testing a single device with the ventilation Thing introduced earlier in Listing 1. The first case will be without using paths to illustrate the state-of-the-art approach and the second case will apply the path vocabulary.

Testing Without Paths Before adding the path vocabulary, one can automatically test a Thing by sending requests described in its TD in a random order, called a test scenario. Combined with the data structure represented in the TD, it is possible to cover every interaction described in the TD of the Thing under test.

We have developed the test architecture in Fig. 8 to test each of the three interaction patterns introduced in Sect. 2. This architecture allows us to systematically test a Thing by using its TD. We run the corresponding interaction pattern's test method (the vertically aligned boxes) for each interaction in the test scenario as follows:

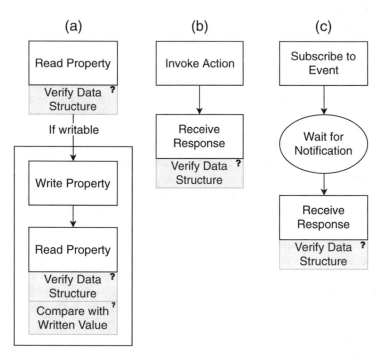

Fig. 8 Architecture of the proposed testing methodology of any interaction of a Thing with a given Thing Description. The yellow boxes (with a ?) symbolize a test that can find either a faulty or correct behavior. The data needed to invoke an action or write to a property is generated using data generation tools

- Property (Fig. 8a): The property value is read and then compared with the structure given in the TD. If the property is writable, a value is generated according to the described data structure and sent to the Thing. The same property is read again to check whether the write request has been successful.
- Action (Fig. 8b): If the action needs input data to execute, the input data is generated and sent to the Thing to invoke the action. Then the response value is compared with the structure given in the TD.
- Event (Fig. 8c): First the event subscription is performed. Once the event is triggered, the value is received and it is compared to the structure given in the TD.

Figure 9 shows an execution trace extract of a test scenario that includes the test of the `rotation` property and the `rotate` action. Here, the Thing under test has interactions that require sequential execution to properly function, but the testing was performed in random order, as the sequence could not be expressed in the TD without paths. This lack of expressiveness makes the test results unreliable. As illustrated in Fig. 9, invoking the `rotate` action and writing to the `rotation` property does not change anything in the system since the `initialize` action has not been invoked before. This is shown as an error because the write operation was not successful, but the real problem is in the order of interactions. This is a problem

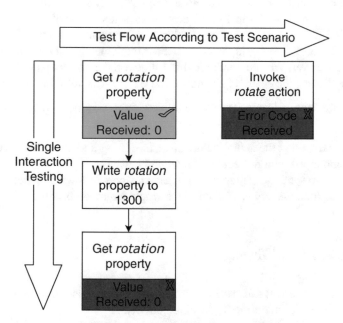

Fig. 9 Illustration of a test path generated from the Thing Description of the industrial ventilator that does not support the path vocabulary. The red boxes (with an X) should symbolize a fault in the ventilator. However, these are the correct responses if the sequential behavior is not respected. The lack of expressiveness in the Thing Description causes this misinterpretation error

found while testing, but the same problem can occur when a consumer Thing (e.g., gateway) is trying to interact with the exposer Thing.

Figure 9 shows two problems originating from the lack of expressiveness regarding sequential behavior:

- The `Rotate` action will *probably* not be used as it is meant to. The only way to do it systematically would be to read a document such as an operation manual and manually write the test scenario.
- Errors in the implementation of the `Rotate` action will never be detected in a systematic way. The `Rotate` action will be used the way it is designed only if the random order of interactions during testing matches the sequential behavior.

Testing with Paths By using the path vocabulary, the randomness of the order of requests can be mitigated. Test scenarios can be generated in a systematic way instead of a random way and thus the actual behavior of the system can be tested. The testing method with vertically ordered boxes of Fig. 8 for testing a single interaction stays the same and only the ordering of the test scenario changes.

By using the path vocabulary, one can automatically generate a test scenario that tests the described sequential behavior. This is illustrated in Fig. 10 where the last test `Get rotation property` is shown to have two outcomes. Normally, there would be only one response. For demonstration purposes, we have illustrated one faulty and one correct response. Compared to the red results (with an X) in Fig. 9, this red result (with an X) in Fig. 10 detects an actual error of the device. In the case of the error outcome, we see a value smaller than the intended one, which can be because of the developer not properly implementing the rotation function of the motor driver. We can conclude that following the correct path allowed us to systematically test the desired behavior of the `write` functionality of the Thing.

There are two advantages of the added expressiveness for testing single Things:

- Test scenarios test the actual behavior of the Thing and show real faults of the Thing under test with respect to its intended behavior.
- More features of the Thing can be tested since following a path describes additional functionality compared to the single interactions alone.

4.2   System Level Testing

In this use case, we will illustrate the testing of the previously introduced temperature controlling system during its development cycle.

As mentioned in Sect. 3.2, it is possible to describe an IoT system in a TD with the path vocabulary. For this specific use case, our gateway/system controller device does not bring any extra functionality and is used only for composing the system. Thus, in its TD, there is no interaction but only paths. It is still the same Thing as described by Listing 3.

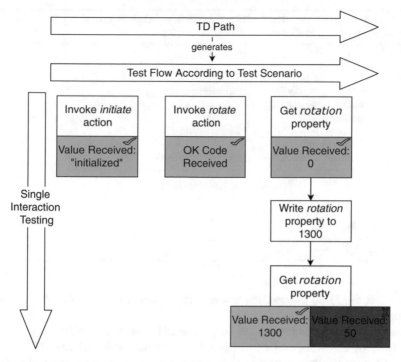

Fig. 10 Illustration of a test path generated from the Thing Description of the industrial ventilator that supports the path vocabulary. Differently from Fig. 9, the correct sequential behavior can be tested and real faults in the system can be identified. The last test case is shown with two possible outcomes, depending on whether the Thing has faults or not, which are both valid test results

All the URIs (lines 10–13) are absolute and they refer to interactions of Things in the system. By following the path named `control` (starting at line 7), the gateway can regulate the temperature of the system. To do so, it gets the temperature value from the temperature sensor, then initializes, and rotates the motor of the ventilator.

Note that paths are composed of interactions as per Definition 3. This means that even though we can use URIs of JSON Pointers that point to paths in a TD, we will decompose a path to its interactions and then include these interactions in a new path. This allows us to keep the same testing logic and not require to modify the definition of a path. The logic stays the same but the system controller will need to fetch the TD that contains the path by using the absolute URI found in the path.

To generalize the testing approach in order to adapt to any TD of a system, and thus to be able to test the complete system, we propose Algorithm 1.

This algorithm allows us to cover the whole system that has arbitrarily many Thing or inter-Thing sequential behaviors. To do so, for every TD of the system (including system controllers) (line 1), it iterates through each path (line 2). In a path, with the listed URIs (line 3), it finds the interaction from every TD using the `findInteraction` function (line 4) and tests the interaction depending on its

Algorithm 1 Algorithm for testing a system of Things based on their TDs that support the path vocabulary

---

1: **for** TD ∈ System **do**
2:      **for** path ∈ TD **do**
3:          **for** uri ∈ path **do**
4:              interactionUnderTest ← findInteraction(uri)
5:              **switch** (interactionType)
6:              **case** property:
7:                  result ← testProperty(interactionUnderTest)
8:              **case** action:
9:                  result ← testAction(interactionUnderTest)
10:             **case** event:
11:                 result ← testEvent(interactionUnderTest)
12:             **end switch**
13:             store TD.path.uri.result
14:         **end for**
15:     **end for**
16: **end for**

---

type (lines 5–12). In the end, the test results are stored to allow diagnostics of the system (line 13).

The test scenario in Fig. 10 can be generated by using Algorithm 1, even if the `initiate`, `rotate`, and `rotation` interactions are in different TDs. Hence, we can automatically generate test scenarios composed of interactions of different Things and test the system composed of several Things.

In this case study, we have demonstrated that paths allow increasing the meaning of test results as well as the quality of tests and hence contribute to improve the testability of IoT systems.

5   Related Work

Thing Description (TD) is a new standard that has resulted from research on Web of Things and Semantic Web technologies, all trying to address the interoperability problem in IoT. As discussed in [12], Web of Things has found application in industry, resulting in its wide adoption and [13] defined the Thing Description standard by using Semantic Web technologies.

For composing an interoperable IoT system, there have been approaches based on marketplaces for IoT devices, such as in [14, 15]. These marketplaces would offer device descriptions for other devices to search for and consequently to use the devices based on their description. For automatically composing a system, a system controller would look for devices it needs, referred to as recipes in [14], from the marketplace and compose the desired system with the devices it finds. However, there is no description of sequential behavior that can link the capabilities of Things in a sequential order.

Mayer et al. [16] introduce a more generic approach where a goal is set using the RESTdec format, such as controlling the temperature, and the system is composed based on this goal. However, the RESTdec format is not human-readable. In addition, Thuluva et al. [14] and Mayer et al. [16] present top-down approaches and the core technology they are using is not standardized as it is with TDs.

In our approach, however, our first contribution is solving the ambiguity of sequential behavior in TDs in a human-readable format on device level by adding the path vocabulary. As a further contribution, we can use it for composing system behavior in a sequential fashion.

Moreover, the path vocabulary is very similar to formal property specification. Hence, in the future, it might enable the application of formal verification methods.

6 Discussion

The black-box testing approach that we have shown in this paper will be used to automate the testing during the standardization activities of the W3C Thing Description standard. As with any W3C standard, the TD standard document has assertions to describe what devices should do and how their TD should be represented. Chapter 8 "Behavioral Assertions" of [1] provides very clear assertions on how a Thing should behave regarding data types of payload and what are the assumptions on protocols.

The existing WoT implementations provided by the community are used to generate a W3C Implementation Report. Together with assertions in other chapters, this report is a mandatory step for the TD standard to be published. The path vocabulary is not yet part of the TD standard, but our testing methodology is equally valid for TDs without paths.

In the current state, the Web of Things Thing Description Implementation Report as seen in Fig. 11 contains 14 implementations that are tested manually for their conformance to the standard. With the increasing number of implementations, applying an automated testing methodology as presented in this paper will be inevitable.

7 Conclusion

In this paper, we introduced a new vocabulary called paths for the Thing Description standard. Using the path vocabulary, we have described sequential behavior of Things in TDs and made it possible to test such behavior automatically, which was not possible in the current standard. We have shown that the same vocabulary can be used for describing a system composed of individual Things without preprogrammed interfaces. Hence, the methodology to test a single Thing was generalized to test systems composed of individual Things. In a case study, we have

ID	Category	Context	Assertion	P	F	N	T
bindings-requirements-scheme	Behavior	Form	Every form in a WoT Thing Description MUST follow the requirements of theProtocol Binding indicated by the URI scheme of its href member.	13	0	3	16
bindings-server-accept	Behavior	(TDConsumer)	Every form in a WoT Thing Description MUST accurately describe requests (including request headers, if present) accepted by the Thing in an interaction.	14	0	3	17
client-data-schema	Behavior	(TDConsumer)	A Thing acting as a Consumer when interacting with another target Thing desciibed in a WoT Thing Description MUST generate data organized according to the data schemas given in the corresponding interactions.	6	0	9	15
client-data-schema-accept-extras	Behavior	(TDConsumer)	A Thing acting as a Consumer when interacting with another Thing MUST accept without error any additional data not described in the data schemas given in the Thing Description of the target Thing.	5	1	9	15
client-data-schema-no-extras	Behavior	(TDConsumer)	A Thing acting asa Consumer when interacting with another Thing MUST NOT generate data not described in the data schemas given in the Thing Description of that Thing.	5	1	9	15
client-uri-template	Behavior	(TD Consumer)	A Thing acting asa Consumer when interacting with another Thing MUST generate URIs according to the URI Templates, base URIs, and form href parameters given in the Thing Description of the target Thing.	3	0	12	15
iana-security-alter	IANA	(TD Consumer)	For this reason, Consumer again SHOULD vet and cache remote contexts before allowing the system to use it.	1	0	6	7
iana-security-execution	IANA	(TD Consumer)	Since WoT Thing Description is intended to be a pure data exchange format for Thing metadata, the serialization SHOULD NOT be passed through a code execution mechanism such as JavaScript's eval() function to be parsed.	5	0	2	7
iana-security-expansion	IANA	(TD Consumer)	Consumers SHOULD treat any TD metadata with due skcpticism.	1	0	4	5
iana-security-remote	IANA	(TD Consumer)	While implementations on resource-constrained devices are expected to perform raw JSON processing (as opposed to JSON-LD processing), implementations in general SHOULD statically cache vetted versions of their supported context extensions and not to follow links to remote contexts.	3	0	5	8
server-data-schema	Behavior	(TD Producer)	A WoT Thing Description MUST accurately describe the data returned and accepted by each interaction.	17	1	3	21
server-data-schema-extras	Behavior	(TD Producer)	A Thing MAY return additional data from an interaction even when such data is not described in the data schemas given in its WoT Thing Description.	7	1	13	21
server-uri-template	Behavior	(TD Producer)	URI Templates, base URIs, and href members in a WoT Thing Description MUST accurately describe the WoT Interface of the Thing.	10	0	1	21

Fig. 11 A snapshot of the Web of Things Thing Description Implementation Report as of 06 May 2019. This is a small section of the report that contains the behavioral assertions. The implementation report is constantly updated with new implementations and it can be found on the official GitHub repository of W3C at https://github.com/w3c/wot-thing-description/blob/master/testing/report.html

shown how testing benefits from the enhanced expressiveness in TDs. Thus, this contribution allows us for the first time using Thing Descriptions to systematically compose and test cyber-physical systems.

References

1. Kaebisch, S., Kamiya, T., McCool, M., & Charpenay, V. (2019). Web of Things (WoT) Thing Description. Candidate recommendation, W3C, https://www.w3.org/TR/2019/CR-wot-thing-description-20190516/.
2. Korkan, E., Kaebisch, S., Kovatsch, M., & Steinhorst, S. (2018). Sequential behavioral modeling for scalable iot devices and systems. In *2018 Forum on Specification Design Languages (FDL)* (pp. 5–16). https://doi.org/10.1109/FDL.2018.8524065.
3. Philips Lighting B.V. (2019). Hue API. https://developers.meethue.com/develop/hue-api/.
4. Bryan, P. C., Zyp, K., & Nottingham, M. (2013). JavaScript Object Notation (JSON) Pointer. RFC 6901. https://doi.org/10.17487/RFC6901. https://rfc-editor.org/rfc/rfc6901.txt.
5. Baldoni, R., Contenti, M., Piergiovanni, S. T., & Virgillito, A. (2003). Modeling publish/-subscribe communication systems: Towards a formal approach. In *Proceedings of the Eighth International Workshop on Object-Oriented Real-Time Dependable Systems (WORDS)*. https://doi.org/10.1109/WORDS.2003.1218097.
6. Koster, M. (2018), Web of Things (WoT) Protocol Binding Templates. Tech. rep., W3C. https://www.w3.org/TR/2018/NOTE-wot-binding-templates-20180405/.
7. Barnett, J., Akolkar, R., Auburn, R., Bodell, M., Burnett, D. C., Carter, J., et al. (2015). State Chart XML (SCXML): State Machine Notation for Control Abstraction. W3C Recommendation, W3C. https://www.w3.org/TR/2015/REC-scxml-20150901/.
8. Bray, T. (2014). The JavaScript Object Notation (JSON) Data Interchange Format. https://rfc-editor.org/rfc/rfc7159.txt. https://doi.org/10.17487/RFC7159.
9. Sporny, M., Lanthaler, M., & Kellogg, G. (2014). JSON-LD 1.0. W3C Recommendation, W3C. http://www.w3.org/TR/2014/REC-json-ld-20140116/.
10. Shelby, Z., Hartke, K., & Bormann, C. (2014). The Constrained Application Protocol (CoAP). https://rfc-editor.org/rfc/rfc7252.txt. https://doi.org/10.17487/RFC7252.
11. The Modbus Organization. (2012). Modbus application protocol specification v1.1b3. http://www.modbus.org/docs/Modbus_Application_Protocol_V1_1b3.pdf.
12. Guinard, D. (2011). http://www.vs.inf.ethz.ch/publ/papers/dguinard-awebof-2011.pdf. PhD thesis, ETH Zurich, Zurich, Switzerland.
13. Charpenay, V., Käbisch, S., & Kosch, H. (2016). Introducing Thing Descriptions and Interactions: An Ontology for the Web of Things. In *Stream Reasoning + Semantic Web technologies for the Internet of Things @ Int. Semantic Web Conference*.
14. Thuluva, A., Bröring, A., Medagoda, G., Don, H., Anicic, D., & Seeger, J. (2017). Recipes for IoT Applications. In *Proceedings of the Seventh International Conference on the Internet of Things*. New York: ACM. https://doi.org/10.1145/3131542.3131553.
15. Bröring, A., Schmid, S., Schindhelm, C. K., Khelil, A., Käbisch, S., Kramer, D., et al. (2017). Enabling IoT Ecosystems through Platform Interoperability. *IEEE Software, 34*(1), https://doi.org/10.1109/MS.2017.2.
16. Mayer, S., Verborgh, R., Kovatsch, M., & Mattern, F. (2016). Smart configuration of smart environments. *IEEE Transactions on Automation Science and Engineering*. https://doi.org/10.1109/TASE.2016.2533321.

Automatic Design of Microfluidic Devices: An Overview of Platforms and Corresponding Design Tasks

Robert Wille, Bing Li, Rolf Drechsler, and Ulf Schlichtmann

1 Introduction

Microfluidic devices provide a more convenient and cost-effective way to conduct biochemical, biological, or medical experiments [29, 58]. Instead of conducting tests manually in a fully equipped lab using expensive lab equipment and human resources, these devices allow to conduct biochemical and medical experiments on a small chip—yielding the so-called *Labs-on-Chips* (LoCs). This requires much smaller sample/reagent volumes and leads to a significantly higher throughput. Examples in which microfluidic devices have successfully been applied include, e.g., PCR [28], protein crystallization [89], sample preparation [3], nanoparticle synthesis [42], drug screening [31], or encapsulation [32, 67].

However, designing the corresponding chips has become a considerably complex task. Depending on the respective platform thousands—or even tens of thousands—of entities and features have to be put together and/or dedicated physical characteristics (e.g., the flow of fluids or the resistance of channels) have to be considered. Despite these challenges, most of the microfluidic devices are still designed manually thus far. This frequently leads to designs that often do not

R. Wille (✉)
Johannes Kepler University Linz, Linz, Austria
e-mail: robert.wille@jku.at

B. Li · U. Schlichtmann
Technical University of Munich, Munich, Germany
e-mail: b.li@tum.de; ulf.schlichtmann@tum.de

R. Drechsler
University of Bremen and DFKI GmbH, Bremen, Germany
e-mail: drechsler@uni-bremen.de

© Springer Nature Switzerland AG 2020
T. J. Kazmierski et al. (eds.), *Languages, Design Methods, and Tools for Electronic System Design*, Lecture Notes in Electrical Engineering 611,
https://doi.org/10.1007/978-3-030-31585-6_4

perfectly work as desired after the first try, but require frequent (costly and time-consuming) iterations.

At the same time, several methods and solutions for the design automation of microfluidic devices have been proposed in the past years. Although they are not that heavily used by the actual stakeholders yet, they provide a starting point for introducing and exploiting EDA methods in the microfluidic domain. However, in order to truly introduce design automation to the microfluidic community, the respective methods need to be much more focused on the actual needs of these stakeholders. Besides other issues, this also requires the resulting tools to be much more accessible and significantly simpler to use.

This requires experts from the design automation community to be familiar with the respective platforms as well as the corresponding design challenges. In this tutorial, we aim for providing an introduction to both issues. To this end, we provide an overview on different microfluidic platforms (including devices based on electrowetting and continuous flows as well as solutions based on a passive routing concept) and the corresponding design tasks. Afterwards, we sketch how to automatically address these design tasks. References are provided to equip the interested reader with comprehensive descriptions for a more in-depth treatment. Overall, this shall provide a starting point for researchers and engineers interested in getting involved in this area.

2 Electrowetting-Based Microfluidic Devices

The first platform considered in this tutorial relies on a discretization of the considered fluids into the so-called *droplets* of picoliter or nanoliter size. This is accomplished by a technique called electrowetting [29, 62] and eventually yields microfluidic devices usually referred to as *Digital Microfluidic Biochips* (DMFBs).

2.1 The Platform

A DMFB is a two-dimensional electrical *grid* controlled by underlying electrodes and their electrical actuations. Using those, an electric field is generated which allows to "hold" discretized portions of fluids, the *droplets*, on a particular cell within the grid. By assigning time-varying voltage values to turn electrodes on and off, droplets can be moved around the grid. This technique, called *electrowetting-on-dielectric* [62], eventually provides a platform on which droplets derived from laboratory fluids such as blood, urine, or corresponding reagents can be exposed to several *operations* such as mixing, heating, or analyzing.

These operations are realized by the so-called *modules* which may be physically built onto the chip or are virtually realized through electrowetting. More precisely, *physical modules* include:

- *Dispensers*: Fluids to be used in the experiment are kept in the so-called *reservoirs*. Whenever required, a sample, i.e., a droplet of the corresponding fluids, is taken from this reservoir and placed onto the grid. For this purpose, *dispensers* for each fluid are physically added next to the outer cells of the grid. For each *type* of fluid considered in the experiment (e.g., blood, urine, reagents), a separate reservoir and, hence, a separate dispenser has to be provided.

- *Sinks*: If droplets are not needed anymore during the execution of an experiment, they shall be removed from the grid (e.g., in order to make room for other droplets and/or operations). For this purpose, similar to dispensers, sinks are added to the outer cells of the grid. Since sinks are used for waste disposal only, no differentiation between types is necessary.
- *Heaters*: Heating samples may be an integral part of an experiment. To provide this operation, heaters can be added to the chip. For this purpose, heating devices are placed below selective cells. Then, droplets occupying this cell can be heated if desired.
- *Detectors*: At the end of an experiment, the properties of the resulting droplet shall usually be examined. For this purpose, respective sensor devices are placed below selective cells. Then, droplets occupying this cell can be analyzed with respect to different characteristics such as color, volume, etc.

While physical modules always require corresponding devices built-in onto the chip, some of the operations can implicitly be realized by the movements of droplets (which in turn is realized through electrowetting as described above). In the following, these modules are called *virtual modules*. Examples include:

- *Mixers*: Mixing fluids (represented by droplets) is an integral part of almost every experiment. Using electrowetting, this can be realized by simply routing the respective droplets to be mixed to the same cell. In order to accelerate diffusion, the newly formed droplet is moved back and forth between several cells.
- *Splitters*: Droplets resulting from mixing operations have twice the size than the input droplets. To reduce them to normal size, they are split up. This can be realized by simultaneously activating cells of the grid that are on the opposite sides of the droplet. Then, the resulting forces split the droplet into two parts.

Overall, modules allow for the realization of various operations to be performed in laboratory experiments. Some of them are available in different fashions with respect to the number of occupied cells and the number of timesteps required for their execution. A list of all available modules (including their implementations) is provided in a *module library*.

Example Figure 1 illustrates the realization of an experiment on a 5×5 grid. In the first timestep, the droplets d_1, d_2, and d_3 are dispensed onto the chip. While the droplets d_1 and d_2 are mixed for 4 timesteps in mixer m_1, droplet d_3 is heated to its desired temperature for 3 timesteps. The heated droplet d_3 and the result of the mixing operation are then mixed for another 7 timesteps. The resulting droplet is eventually analyzed by the detector in timesteps 15–21. As can be seen, different fashions of modules are applied for the mixing operation. The first mixer required a 2×2 subgrid and 5 timesteps, while the second one occupied a 1×3 subgrid over 7 timesteps. □

Fig. 1 An experiment
conducted on a DMFB

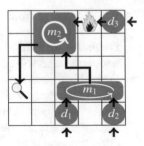

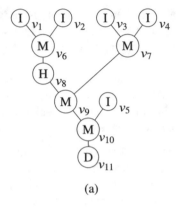

RESOURCE	AREA	TIME
Dispenser (I)	–	1
Mixer$_1$	2×3	7
Mixer$_2$	3×3	5
Heater	1×1	10
Detector	1×1	15

(b)

AREA	TIME
5×5	50

(c)

(a)

Fig. 2 Specification of an experiment. (**a**) Sequencing graph. (**b**) Module library. (**c**) Constraints

2.2 The Design Process

To design DMFBs, several automatic synthesis methods have been proposed in the recent past. These methods require a specification of the experiment to be realized as well as the resources available for this purpose. More precisely, the following input is usually provided:

- A *sequencing graph* which specifies the experiment to be realized by the involved fluids (in terms of droplets) as well as the respective steps (in terms of operations) and their dependencies of execution,
- A *module library* providing the available modules which can be used in order to realize the respective operations given in the sequencing graph, and
- Additional *constraints*, e.g., on the size of the grid on which the experiment shall be conducted or the maximal duration of the experiment.

Example Figure 2 provides a specification of an experiment to be realized on a DMFB. The sequencing graph in Fig. 2a defines the dispensing operations (v_1 to v_5) and their successors. The module library in Fig. 2b lists the modules available to realize those operations. Additionally, constraints as shown in Fig. 2c state that the entire experiment is to be conducted onto a 5×5 grid taking at most 50 timesteps.

□

Having these inputs, the following design questions need to be addressed:

- Which modules shall be applied in order to realize an operation?
- When (at what timesteps) shall each operation be conducted?
- Where (on which cells or subgrid) shall each operation be conducted?
- How shall the respective droplets be routed towards their destination?
- What pins/cells need to be actuated in order to realize the respective operations and routings onto the grid?

All these questions eventually represent typical system design tasks such as *binding*, *scheduling* [27, 64], *placement* [1, 7, 63, 88], and *routing* [41, 46, 65, 85, 87], respectively, for which dedicated DMFB-related solutions have been proposed as given in the references. In addition, the pin-actuation problem is addressed in works such as [34, 45]. Finally, initial approaches for a *one-pass design flow* have been introduced in [47, 84]—aiming for conducting all these tasks in a single and integrated process.

Recently, also extensions of DMFBs are considered in which droplets are not actuated by single electrodes but a sea-of-micro-electrodes—yielding the so-called *micro-electrode-dot-array biochips* (MEDA biochips, see, e.g., [50, 78, 79]). This additionally allows for a much greater flexibility, e.g., through allowing droplets of rather arbitrary sizes, diagonal movements, etc. With the emerge of this extended platform, also correspondingly adjusted design methods have been proposed, e.g., in [43, 44, 52].

3 Flow-Based Microfluidic Devices

The second platform covered by this tutorial is composed of microchannels and microvalves, which are, respectively, called channels and valves for simplicity. A channel is etched on a substrate to conduct fluid samples between devices. The movement of fluid samples is coordinated by valves, whose states are controlled by air pressure patterns [57, 70, 77]. Since fluid segments instead of droplets are manipulated on such a platform, it is thus referred to as *flow-based microfluidic biochips*.

3.1 The Platform

A flow-based microfluidic biochip has a structure with two layers. Flow channels are itched on silicon/glass substrates or made from dimethylsiloxane using soft lithography [59] to transport fluid samples and reagents between devices. Above flow channels, control channels are used to deliver air pressure to the crossing points between flow channels and control channels. Both, flow and control channels, are made from elastic materials, so that air pressure in a control channel extends it

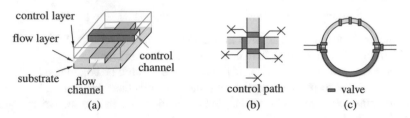

Fig. 3 Components in flow-based microfluidic devices. (**a**) Valve structure. (**b**) Switch. (**c**) Mixer

and, thus, squeezes the flow channel underneath. Consequently, the movement of the fluid sample or reagent in the flow channel is blocked. After the pressure in the control channel is removed, fluid transportation in the flow channel can be resumed. Consequently, valves are constructed at these crossing points, as illustrated in Fig. 3a.

Valves are the basic flow control components in a flow-based microfluidic biochip. When multiple transportation channels intersect with each other, only one channel can be used simultaneously. To avoid fluid contamination, valves need to be built at the intersections of flow channels to direct the flow transportation, in fact forming switching as shown in Fig. 3b. At a given moment, only two of the four valves in the switch open to allow fluid to pass. This transportation control can be configured dynamically by changing the states of these valves with respect to the requirements of the application, so that complex biochemical assays can be executed by simple biochips with time multiplexing.

Using valves, more complex devices such as mixers can also be implemented. For example, in Fig. 3c, three valves are constructed along a circular channel at the top. If these valves are switched alternately with a pattern 101, 100, 110, 010, 011, 001, where 1 means the valve is open and 0 means the valve is closed, a flow along the circular channel can be generated—emulating the peristaltic effect for mixing fluid samples and reagents [59].

In a flow-based biochip, channels are used to transport fluid samples. If a fluid sample resides in a channel segment instead of being moved, this sample can be considered as stored inside this channel segment. This feature is very useful since it is easy to implement the storage function anywhere inside a flow-based biochip. When multiple channels are arranged side-by-side and multiplexing is implemented at the input and the output of these channels, a dedicated storage unit can also be implemented.

Example Figure 4 shows a flow-based microfluidic biochip with a mixer at the top and a dedicated storage unit at the bottom. At a given moment, the valves at the input and the output of the storage unit only allow one fluid sample to be moved into or out of the storage unit—similar to memory blocks in electronic systems. The mixer can be used to mix fluid samples entering the chip from the two input ports. Intermediate results can be saved in the storage unit temporarily and fetched later for further processing. Through the oil port, a flow path can be constructed to push

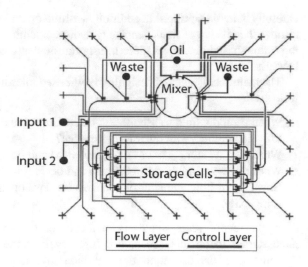

Fig. 4 Flow-based biochip w/ mixer and storage unit [2]

fluid samples between devices. The waste ports are used to discard fluid samples that are of no use anymore. □

Similar to DMFBs, further dedicated devices such as heaters, filters, and detectors can also be constructed in a flow-based biochip to provide specific functions. As a result, a flow-based biochip can be considered as a channel network connecting dedicated devices. Unlike in DMFBs, all these devices are dedicated and operations must be executed by the corresponding devices at given locations. Intermediate result must be transported between these devices through the channel network to execute complex biochemical applications.

3.2 The Design Process

Designing flow-based microfluidic biochips is similar to designing DMFBs. A sequencing graph describing the experiment protocol as shown in Fig. 2a is used to define what operations need to be executed and how their results need to be transported. Furthermore, the devices available to a flow-based biochip can also be described as a module library, similar to Fig. 2b. The difference is that the areas of these devices do not matter so much as in DMFBs, because these devices are pre-built on the chip instead of being formed on the chip on-the-fly.

Since devices in flow-based biochips are fixed at given locations, the results from these devices should be moved between them through the channel network that connects the devices. This is the major difference between a flow-based biochip and a DMFB, because the latter allows the locations of devices to be moved so that fluid transportation is more flexible. When multiple fluid samples are moved across a channel network in a flow-based biochip, fluid transportation needs to be arranged carefully to avoid conflicts. In addition, washing operations need to be performed to remove the residue of fluid samples to avoid contamination. Consequently, designing a flow-based biochip is more transportation-centered compared to designing a DMFBs.

The major challenges of designing flow-based microfluidic biochips are listed as follows:

* When should a fluid transportation be conducted and when should it be stopped to avoid conflicts with other fluid samples?
* Where should storage units be implemented and how large should they be?
* When and how should flow channels and devices be washed?
* How should flow and control channels be developed together to reduce design complexity.

In the recent years, synthesis methods for flow-based microfluidic biochips have started to be introduced. For high-level synthesis, the workflows in [60, 71] minimize the execution time of bioassays and valve switching activities, respectively. In addition, a distributed storage system is proposed in [56, 74] to improve transportation efficiency. Moreover, washing is implemented in [38, 39] to avoid contamination. For physical design, the placement of devices and routing of channels in flow-based biochips are dealt with simultaneously in [80] and formulated as an SAT problem in [26] to achieve a close-to-optimal result.

Control logic synthesis is investigated in [61] and the method in [35] minimizes pressure propagation delay to reduce the response time of valves. Switching patterns of valves are examined in [81, 82] to reduce the largest number of valve switching activities to improve the reliability of valves, and length-matching is incorporated in control channel routing in [86]. Flow-layer, control-layer, and valve switching are considered together in [75, 76] to simplify the overall design complexity.

Fault models and an ATPG-based test strategy for flow-based biochips are proposed in [36, 40] to deal with manufacturing defects. Design-for-testability and defect diagnosis are further addressed in [33, 37, 55].

To provide better reliability and flexibility, *Programmable Microfluidic Devices* (PMDs) [14] have been explored in [72, 73]. Channel crossing on a general array architecture is avoided in [49] and valve control sequences are arranged carefully for such a chip in [25]. Test generation is introduced in [54] to improve test efficiency.

4   Passive Routing Concepts for Microfluidic Devices

Both platforms reviewed above rely on an *active* control method realized either by actuations of electrodes or dedicated valves—resulting in rather costly and error-prone solutions. As an alternative to that, another platform recently got investigated which entirely relies on a *passive routing concept*. This concept has been applied, e.g., in *Networked Labs-on-Chips* [10] and *Hydrodynamic Controlled Microfluidic Networks* [9].

Fig. 5 Bifurcation

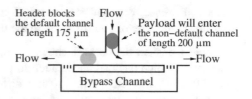

Header blocks
the default channel
of length 175 μm

Flow

Payload will enter
the non–default channel
of length 200 μm

Flow ◄—   —►Flow

Bypass Channel

4.1 The Platform

Passive routing concepts can be realized on top of two-phase flow microfluidics, where the, respectively, considered droplets flow in an immiscible continuous flow inside closed channels. Pumps generating the continuous flow eventually distribute this flow among the network, which may consist of a set of modules executing unit operations. By this, the, respectively, injected droplets will be passed through a particular path of modules—executing operations such as mixing, splitting, delaying, incubating, detecting, or heating [48, 53, 66, 68, 83] and, hence, realizing the desired medical/biochemical experiment.

In order to explicitly route droplets along the desired paths (without using active controls based on electrodes or valves), the so-called bifurcations and corresponding hydrodynamic forces are exploited. More precisely, a *bifurcation* as shown in Fig. 5 yields different volumetric flow rates in its successor channels which depend on the respective geometries of those channels. For example, the smaller the diameter and/or the longer the channel, the higher the resistance and viscosity of the continuous phase.[1] Because of that, a *single* droplet arriving at a bifurcation will always flow along the successor with the lower fluidic resistance (called the *default* successor) [8, 16]. However, since droplets themselves increase the resistance of a channel (e.g., through their viscosities, droplet size, and geometry as studied, e.g., in [4, 15, 17]), they temporarily block the default successor for the following droplets—allowing a following droplet to take a different path (as observed and/or simulated, e.g., in [9–13, 51]).

These concepts of default successors at bifurcations and the possibility to block them with other droplets allow to realize arbitrary paths through a microfluidic network. More precisely, if the actually considered droplet (called *payload* droplet) is supposed to take a non-default successor at any bifurcation in the network, it has to make sure that another droplet (called *header* droplet) arrives before and blocks the default successor. This is accordingly sketched in Fig. 5, where the blue droplet (the header) blocks the path so that the green droplet (the payload) takes the intended path. Overall, this allows to *passively* route payloads through different

---

[1]Note that a *bypass* channel [8] connects the endpoints of the two successor channels. This bypass cannot be entered by any droplet and is used to make the droplet routing only dependent on the resistances of the successors (and not the entire network).

paths and, hence, different sequences of modules can be executed without any additional hardware or control logic on the device.

Example Consider the network shown in Fig. 6. Here, a pump produces a continuous flow in which payload and header droplets are injected. Then, the droplets can take different paths and, by this, realize different experiments. For example, if just a single payload droplet is injected, only default paths are taken, i.e., the modules mixing, heating, and incubating are executed. If additionally a header is injected at a particular time so that the channel c_4 is blocked when the payload arrives at the second bifurcation, a path of the payload is realized in which the heating step is skipped.   □

In order to avoid that operations of modules are executed on headers, the modules are shielded by a droplet by size sorter [69]. A sorter steers payloads towards the module and forwards headers. Therefore, the sorter uses the different droplet sizes (i.e., droplet volumes) of headers and payloads. Finally, the network contains bifurcations allowing droplets to take multiple paths and, by this, to realize different experiments on a payload. Whether a path is implemented by the default or by the non-default successor channel is also defined by the network.

4.2   The Design Process

Exploiting this routing concept requires a very dedicated and sensitive design as just small differences, e.g., in some channel lengths may change the hydrodynamic forces within the network and, hence, change the behavior of the microfluidic device. Accordingly, the following steps shall be conducted in order to guarantee a correct design.

First, a proper architecture needs to be defined. This strongly depends on the given set of operations to be executed and their corresponding order. In order to allow for a cost-effective architecture, operations can be re-used for different experiments. For example, the experiments shown in Fig. 7a–c can all be realized by an architecture sketched in Fig. 7d. A method automatically determining a suitable method has been proposed in [23].

The resulting architecture can directly be mapped to a structure as shown before in Fig. 6. However, it remains to be defined how to properly dimension the used channels. This constitutes a significant challenge since the dimensions of the channels significantly affect the flow of the droplets. In order to aid designers in this task, methods proposed in [24] allow for automatically determining and validating corresponding dimensions.

Then, payload and header droplets need to be injected into the network at dedicated times. This requires the determination of dedicated droplet injection sequences which make sure that the header droplets arrive in bifurcations at exactly the time when they are supposed to block a default channel. For ring networks as, e.g., proposed in [6, 9, 11, 30, 51], the injection time of the header and payload

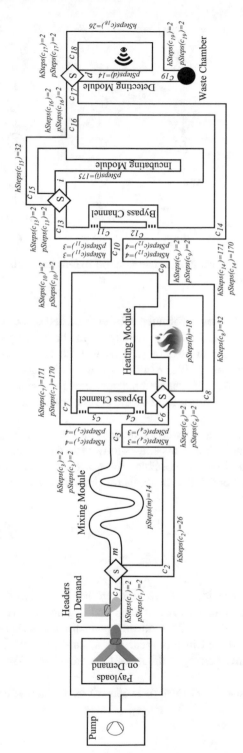

Fig. 6 Microfluidic network supporting passive droplet routing

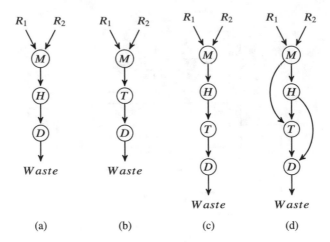

Fig. 7 Given experiments and resulting architecture. (**a**) Experiment 1. (**b**) Experiment 2. (**c**) Experiment 3. (**d**) Res. Arch.

droplets can be calculated by a formula. However, if dedicated architectures are employed as shown in Fig. 7d, more elaborated methods are required. To this end, a discrete model as proposed in [21] as well as corresponding automatic search methods as proposed in [22] can be utilized. This may even unveil that a corresponding droplet sequences cannot be determined for a given architecture and dimensioning which makes verification of the corresponding devices an important design step [22].

Finally, the resulting design as well as the determined droplet sequences shall be simulated prior to its fabrication. This allows to validate the correct execution of the design and pin-points designers to possible problems before physically realizing the obtained designs. To this end, initial methods for simulation are available, e.g., in [5, 18–20].

5   Conclusions

This tutorial summary provided an overview on different microfluidic devices as well as corresponding challenges researchers and engineers have to tackle when designing them. This shall provide a starting point for researchers and engineers interested in getting involved in this area. For a more in-depth treatment of the respective issues, we are referring to the references provided below.

Acknowledgements We sincerely thank all co-authors and collaborators who worked with us in the past in this exciting area.

References

1. Alistar, M., Pop, P., & Madsen, J. (2013). Operation placement for application-specific digital microfluidic biochips. In *2013 Symposium on Design, Test, Integration and Packaging of MEMS/MOEMS (DTIP)* (pp. 1–6). Piscataway: IEEE.
2. Amin, N., Thies, W., & Amarasinghe, S. P. (2009). Computer-aided design for microfluidic chips based on multilayer soft lithography. In *Proceedings of the International Conference on Computer Design* (pp. 2–9)
3. Bhattacharjee, S., Wille, R., Huang, J. D., & Bhattacharya, B. B. (2018). Storage-aware sample preparation using flow-based microfluidic labs-on-chip. In *Design, Automation and Test in Europe* (pp. 1399–1404).
4. Biral, A., & Zanella, A. (2013). Introducing purely hydrodynamic networking functionalities into microfluidic systems. *Journal of Nano Communication Networks, 4*(4), 205–215.
5. Biral, A., Zordan, D., & Zanella, A. (2015). Modeling, simulation and experimentation of droplet-based microfluidic networks. *IEEE Transactions on Molecular, Biological, and Multi-scale Communications, 1*(2), 122–134.
6. Castorina, G., Reno, M., Galluccio, L., & Lombardo, A. (2017). Microfluidic networking: Switching multidroplet frames to improve signaling overhead. *Journal of Nano Communication Networks, 14*, 48–59.
7. Chen, Y. H., Hsu, C. L., Tsai, L. C., Huang, T. W., & Ho, T. Y. (2013). A reliability-oriented placement algorithm for reconfigurable digital microfluidic biochips using 3-D deferred decision making technique. *IEEE Transactions on Computer-Aided Design of Integrated Circuits and Systems, 32*(8), 1151–1162.
8. Cristobal, G., Benoit, J. P., Joanicot, M., & Ajdari, A. (2006). Microfluidic bypass for efficient passive regulation of droplet traffic at a junction. *Applied Physics Letters, 89*(3), 34104–34104.
9. De Leo, E., Donvito, L., Galluccio, L., Lombardo, A., Morabito, G., & Zanoli, L. M. (2013). Communications and switching in microfluidic systems: Pure hydrodynamic control for networking Labs-on-a-Chip. *IEEE Transactions on Communications, 61*(11), 4663–4677.
10. De Leo, E., Galluccio, L., Lombardo, A., & Morabito, G. (2012). Networked labs-on-a-chip (NLoC): Introducing networking technologies in microfluidic systems. *Journal of Nano Communication Networks, 3*(4), 217–228.
11. Donvito, L., Galluccio, L., Lombardo, A., & Morabito, G. (2013). Microfluidic networks: Design and simulation of pure hydrodynamic switching and medium access control. *Journal of Nano Communication Networks, 4*(4), 164–171.
12. Donvito, L., Galluccio, L., Lombardo, A., & Morabito, G. (2014). On the assessment of microfluidic switching capabilities in NLoC networks. In *International Conference on Nanoscale Computing and Communication* (p. 19).
13. Donvito, L., Galluccio, L., Lombardo, A., & Morabito, G. (2015). μ-NET: A network for molecular biology applications in microfluidic chips. *IEEE/ACM Transactions on Networking, 24*(4), 2525–2538.
14. Fidalgo, L. M., & Maerkl, S. J. (2011). A software-programmable microfluidic device for automated biology. *Lab on a Chip, 11*, 1612–1619.
15. Fuerstman, M. J., Lai, A., Thurlow, M. E., Shevkoplyas, S. S., Stone, H. A., & Whitesides, G. M. (2007). The pressure drop along rectangular microchannels containing bubbles. *Journal on Lab on a Chip, 7*(11), 1479–1489.
16. Glawdel, T., Elbuken, C., & Ren, C. (2011). Passive droplet trafficking at microfluidic junctions under geometric and flow asymmetries. *Journal on Lab on a Chip, 11*(22), 3774–3784.
17. Glawdel, T., & Ren, C. L. (2012). Global network design for robust operation of microfluidic droplet generators with pressure-driven flow. *Journal of Microfluidics and Nanofluidics, 13*(3), 469–480.
18. Gleichmann, N., Malsch, D., Horbert, P., & Henkel, T. (2015). Toward microfluidic design automation: A new system simulation toolkit for the in silico evaluation of droplet-based lab-on-a-chip systems. *Journal of Microfluidics and Nanofluidics, 18*(5–6), 1095–1105.

19. Grimmer, A., Chen, X., Hamidovic, M., Haselmayr, W., Ren, C. L., & Wille, R. (2018). Simulation before fabrication: A case study on the utilization of simulators for the design of droplet microfluidic networks. *RSC Advances, 8*(60), 34733–34742.
20. Grimmer, A., Hamidovic, M., Haselmayr, W., & Wille, R. (2018). Advanced simulation of droplet microfluidics. *Journal on Emerging Technologies in Computing Systems, 15*(3), Article no. 26.
21. Grimmer, A., Haselmayr, W., Springer, A., & Wille, R. (2017). A discrete model for Networked Labs-on-Chips: Linking the physical world to design automation. In *Design Automation Conference* (pp. 50:1–50:6).
22. Grimmer, A., Haselmayr, W., Springer, A., & Wille, R. (2017). Verification of Networked Labs-on-Chip architectures. In *Design, Automation and Test in Europe* (pp. 1679–1684).
23. Grimmer, A., Haselmayr, W., Springer, A., & Wille, R. (2018). Design of application-specific architectures for Networked Labs-on-Chips. *IEEE Transactions on Computer-Aided Design of Integrated Circuits and Systems, 37*(1), 193–202.
24. Grimmer, A., Haselmayr, W., & Wille, R. (2018). Automated dimensioning of networked labs-on-chip. *IEEE Transactions on Computer-Aided Design of Integrated Circuits and Systems, 38*(7), 1216–1225.
25. Grimmer, A., Klepic, B., Ho, T. Y., & Wille, R. (2018). Sound valve-control for programmable microfluidic devices. In *Proceedings of the Asia and South Pacific Design and Automation Conference.*
26. Grimmer, A., Wang, Q., Yao, H., Ho, T. Y., & Wille, R. (2017). Close-to-optimal placement and routing for continuous-flow microfluidic biochips. In *Proceedings of the Asia and South Pacific Design and Automation Conference* (pp. 530–535).
27. Grissom, D., & Brisk, P. (2012). Path scheduling on digital microfluidic biochips. In *Proceedings of the 49th Annual Design Automation Conference* (pp. 26–35). New York: ACM.
28. Guttenberg, Z., Müller, H., Habermüller, H., Geisbauer, A., Pipper, J., Felbel, J., et al. (2005). Planar chip device for PCR and hybridization with surface acoustic wave pump. *Journal on Lab on a Chip, 5*(3), 308–317.
29. Haeberle, S., & Zengerle, R. (2007). Microfluidic platforms for Lab-on-a-Chip applications. *Journal on Lab on a Chip, 7*, 1094–1110.
30. Haselmayr, W., Biral, A., Grimmer, A., Zanella, A., Springer, A., & Wille, R. (2017). Addressing multiple nodes in Networked Labs-on-Chips without payload re-injection. In *International Conference on Communications.*
31. Haselmayr, W., Hamidović, M., Grimmer, A., & Wille, R. (2018). Fast and flexible drug screening using a pure hydrodynamic droplet control. In *European Conference on Microfluidics.*
32. He, M., Edgar, J. S., Jeffries, G. D., Lorenz, R. M., Shelby, J. P., & Chiu, D. T. (2005). Selective encapsulation of single cells and subcellular organelles into picoliter-and femtoliter-volume droplets. *Journal of Analytical Chemistry, 77*(6), 1539–1544.
33. Hu, K., Bhattacharya, B. B., & Chakrabarty, K. (2015). Fault diagnosis for flow-based microfluidic biochips. In *Proceedings of the VLSI Test Symposium* (pp. 1–6).
34. Hu, K., Dinh, T., Ho, T. Y., & Chakrabarty, K. (2016). Control-layer routing and control-pin minimization for flow-based microfluidic biochips. *IEEE Transactions on Computer-Aided Design of Integrated Circuits and Systems, 36*(1), 55–68.
35. Hu, K., Dinh, T. A., Ho, T. Y., & Chakrabarty, K. (2017). Control-layer routing and control-pin minimization for flow-based microfluidic biochips. *IEEE Transactions on Computer-Aided Design of Integrated Circuits and Systems, 36*(1), 55–68.
36. Hu, K., Ho, T. Y., & Chakrabarty, K. (2013). Testing of flow-based microfluidic biochips. In *Proceedings of the VLSI Test Symposium* (pp. 1–6).
37. Hu, K., Ho, T. Y., & Chakrabarty, K. (2014). Test generation and design-for-testability for flow-based mVLSI microfluidic biochips. In *Proceedings of the VLSI Test Symposium* (pp. 97–102).
38. Hu, K., Ho, T. Y., & Chakrabarty, K. (2014). Wash optimization for cross-contamination removal in flow-based microfluidic biochips. In *Proceedings of the Asia and South Pacific Design and Automation Conference* (pp. 244–249).

39. Hu, K., Ho, T. Y., & Chakrabarty, K. (2016). Wash optimization and analysis for cross-contamination removal under physical constraints in flow-based microfluidic biochips. *IEEE Transactions on CAD of Integrated Circuits and Systems, 35*(4), 559–572.
40. Hu, K., Yu, F., Ho, T. Y., & Chakrabarty, K. (2014). Testing of flow-based microfluidic biochips: Fault modeling, test generation, and experimental demonstration. *IEEE Transactions on Computer-Aided Design of Integrated Circuits and Systems, 33*(10), 1463–1475.
41. Huang, T. W., & Ho, T. Y. (2009). A fast routability- and performance-driven droplet routing algorithm for digital microfluidic biochips. In *International Conference on Computer Design* (pp. 445–450). Piscataway: IEEE.
42. Hung, L. H., Choi, K. M., Tseng, W. Y., Tan, Y. C., Shea, K. J., & Lee, A. P. (2006). Alternating droplet generation and controlled dynamic droplet fusion in microfluidic device for CdS nanoparticle synthesis. *Journal on Lab on a Chip, 6*(2), 174–178.
43. Keszocze, O., Ibrahim, M., Wille, R., Chakrabarty, K., & Drechsler, R. (2018). Exact synthesis of biomolecular protocols for multiple sample pathways on digital microfluidic biochips. In *Conference on VLSI Design* (pp. 121–126).
44. Keszocze, O., Li, Z., Grimmer, A., Wille, R., Chakrabarty, K., & Drechsler, R. (2017). Exact routing for micro-electrode-dot-array digital microfluidic biochips. In *Asia and South Pacific Design Automation Conference*.
45. Keszocze, O., Wille, R., Chakrabarty, K., & Drechsler, R. (2015). A general and exact routing methodology for digital microfluidic biochips. In *International Conference on Computer-Aided Design* (pp. 874–881).
46. Keszocze, O., Wille, R., & Drechsler, R. (2014). Exact routing for digital microfluidic biochips with temporary blockages. In *International Conference on Computer-Aided Design* (pp. 405–410).
47. Keszocze, O., Wille, R., Ho, T. Y., & Drechsler, R. (2014). Exact one-pass synthesis of digital microfluidic biochips. In *Design Automation Conference* (pp. 1–6).
48. Köhler, J., Henkel, T., Grodrian, A., Kirner, T., Roth, M., Martin, K., et al. (2004). Digital reaction technology by micro segmented flow-components, concepts and applications. *Chemical Engineering Journal, 101*(1), 201–216.
49. Lai, G. R., Lin, C. Y., & Ho, T. Y. (2018). Pump-aware flow routing algorithm for programmable microfluidic devices. In *Proceedings of the Design, Automation, and Test Europe Conference*.
50. Lai, K., Yang, Y.-T., Lee, C.-Y. (2015). An intelligent digital microfluidic processor for biomedical detection. *Journal of Signal Processing Systems, 78*, 85–93.
51. Leo, E. D., Donvito, L., Galluccio, L., Lombardo, A., Morabito, G., & Zanoli, L. M. (2013). Design and assessment of a pure hydrodynamic microfluidic switch. In *International Conference on Communications* (pp. 3165–3169).
52. Li, Z., Lai, K. Y. T., Yu, P. H., Ho, T. Y., Chakrabarty, K., & Lee, C. Y. (2016). High-level synthesis for micro-electrode-dot-array digital microfluidic biochips. In *Design Automation Conference* (p. 146).
53. Link, D., Anna, S. L., Weitz, D., & Stone, H. (2004). Geometrically mediated breakup of drops in microfluidic devices. *Physical Review Letters, 92*(5), 054503.
54. Liu, C., Li, B., Bhattacharya, B. B., Chakrabarty, K., Ho, T. Y., & Schlichtmann, U. (2017). Testing microfluidic fully programmable valve arrays (FPVAs). In *Proceedings of the Design, Automation, and Test Europe Conference* (pp. 91–96).
55. Liu, C., Li, B., Ho, T. Y., Chakrabarty, K., & Schlichtmann, U. (2018). Design-for-testability for continuous-flow microfluidic biochips. In *Proceedings of the Design Automation Conference*.
56. Liu, C., Li, B., Yao, H., Pop, P., Ho, T. Y., & Schlichtmann, U. (2017). Transport or store? Synthesizing flow-based microfluidic biochips using distributed channel storage. In *Proceedings of the Design Automation Conference* (pp. 49:1–49:6).
57. Manz, A., Graber, N., & Widmer, H. M. (1990). Miniaturized total chemical analysis systems: A novel concept for chemical sensing. *Sensors and Actuators B: Chemical, 1*(1–6), 244–248.

58. Mark, D., Haeberle, S., Roth, G., von Stetten, F., & Zengerle, R. (2010). Microfluidic Lab-on-a-Chip platforms: Requirements, characteristics and applications. *Journal of Chemical Society Reviews, 39*(3), 1153–1182.
59. Melin, J., & Quake, S. (2007). Microfluidic large-scale integration: the evolution of design rules for biological automation. *Annual Review of Biophysics and Biomolecular Structure, 36*, 213–231.
60. Minhass, W. H., Pop, P., Madsen, J., & Blaga, F. S. (2012). Architectural synthesis of flow-based microfluidic large-scale integration biochips. In *Proceedings of the International Conference on Compilers, Architecture, and Synthesis for Embedded Systems* (pp. 181–190).
61. Minhass, W. H., Pop, P., Madsen, J., & Ho, T. Y. (2013). Control synthesis for the flow-based microfluidic large-scale integration biochips. In *Proceedings of the Asia and South Pacific Design and Automation Conference* (pp. 205–212).
62. Pollack, M. G., Shenderov, A. D., & Fair, R. B. (2002). Electrowetting-based actuation of droplets for integrated microfluidics. *Journal on Lab on a Chip, 2*(2), 96–101.
63. Su, F., & Chakrabarty, K. (2006). Module placement for fault-tolerant microfluidics-based biochips. *ACM TODAES, 11*(3), 682–710.
64. Su, F., & Chakrabarty, K. (2008). High-level synthesis of digital microfluidic biochips. *ACM JETC, 3*(4), 1:1–1:32. https://doi.org/10.1145/1324177.1324178.
65. Su, F., Hwang, W., & Chakrabarty, K. (2006). Droplet routing in the synthesis of digital microfluidic biochips. In *Design, Automation and Test in Europe* (Vol. 1, pp. 1–6). Piscataway: IEEE.
66. Tan, Y. C., Fisher, J. S., Lee, A. I., Cristini, V., & Lee, A. P. (2004). Design of microfluidic channel geometries for the control of droplet volume, chemical concentration, and sorting. *Journal on Lab on a Chip, 4*(4), 292–298.
67. Tan, Y. C., Hettiarachchi, K., Siu, M., Pan, Y. R., & Lee, A. P. (2006). Controlled microfluidic encapsulation of cells, proteins, and microbeads in lipid vesicles. *Journal of the American Chemical Society, 128*(17), 5656–5658.
68. Tan, Y. C., Ho, Y. L., & Lee, A. P. (2007). Droplet coalescence by geometrically mediated flow in microfluidic channels. *Journal of Microfluidics and Nanofluidics, 3*(4), 495–499.
69. Tan, Y. C., Ho, Y. L., & Lee, A. (2008). Microfluidic sorting of droplets by size. *Journal of Microfluidics and Nanofluidics, 4*(4), 343–348.
70. Thorsen, T., Maerkl, S. J., & Quake, S. R. (2002). Microfluidic large-scale integration. *Science, 298*(5593), 580–584.
71. Tseng, K. H., You, S. C., Liou, J. Y., & Ho, T. Y. (2013). A top-down synthesis methodology for flow-based microfluidic biochips considering valve-switching minimization. In *Proceedings of the International symposium on Physical Design* (pp. 123–129).
72. Tseng, T. M., Li, B., Ho, T. Y., & Schlichtmann, U. (2015). Reliability-aware synthesis for flow-based microfluidic biochips by dynamic-device mapping. In *Proceedings of the Design Automation Conference* (pp. 141:1–141:6).
73. Tseng, T. M., Li, B., Li, M., Ho, T. Y., & Schlichtmann, U. (2016). Reliability-aware synthesis with dynamic device mapping and fluid routing for flow-based microfluidic biochips. *IEEE Transactions on Computer-Aided Design of Integrated Circuits and Systems, 35*(12), 1981–1994.
74. Tseng, T. M., Li, B., Schlichtmann, U., & Ho, T. Y. (2015). Storage and caching: Synthesis of flow-based microfluidic biochips. *IEEE Design and Test, 32*(6), 69–75.
75. Tseng, T. M., Li, M., Freitas, D. N., McAuley, T., Li, B., Ho, T. Y., et al. (2018). Columba 2.0: A co-layout synthesis tool for continuous-flow microfluidic biochips. *IEEE Transactions on Computer-Aided Design of Integrated Circuits and Systems, 37*(8), 1588–1601.
76. Tseng, T. M., Li, M., Li, B., Ho, T. Y., & Schlichtmann, U. (2016). Columba: Co-layout synthesis for continuous-flow microfluidic biochips. In *Proceedings of the Design Automation Conference* (pp. 147:1–147:6).
77. Verpoorte, E., & Rooij, N. F. D. (2003). Microfluidics meets MEMS. *Proceedings of the IEEE, 91*(6), 930–953.

78. Wang, G., Teng, D., & Fan, S. K.: Digital microfluidic operations on micro-electrode dot array architecture. *IET Nanobiotechnology, 5*(4), 152–160 (2011).
79. Wang, G., Teng, D., Lai, Y. T., Lu, Y. W., Ho, Y., & Lee, C. Y. (2013). Field-programmable lab-on-a-chip based on microelectrode dot array architecture. *IET Nanobiotechnology, 8*, 163–171.
80. Wang, Q., Ru, Y., Yao, H., Ho, T. Y., & Cai, Y. (2016). Sequence-pair-based placement and routing for flow-based microfluidic biochips. In *Proceedings of the Asia and South Pacific Design and Automation Conference* (pp. 587–592).
81. Wang, Q., Xu, Y., Zuo, S., Yao, H., Ho, T. Y., Li, B., et al. (2017). Pressure-aware control layer optimization for flow-based microfluidic biochips. *IEEE Transactions on Biomedical Circuits and Systems, 11*(6), 1488–1499.
82. Wang, Q., Zuo, S., Yao, H., Ho, T. Y., Li, B., Schlichtmann, U., et al. (2017). Hamming-distance-based valve-switching optimization for control-layer multiplexing in flow-based microfluidic biochips. In *Proceedings of the Asia and South Pacific Design and Automation Conference* (pp. 524–529).
83. Wang, W., Yang, C., & Li, C. M. (2009). On-demand microfluidic droplet trapping and fusion for on-chip static droplet assays. *Journal on Lab on a Chip, 9*(11), 1504–1506.
84. Wille, R., Keszocze, O., Drechsler, R., Boehnisch, T., & Kroker, A. (2015). Scalable one-pass synthesis for digital microfluidic biochips. *Journal on Design and Test, 32*(6), 41–50.
85. Xu, T., & Chakrabarty, K. (2007). Integrated droplet routing in the synthesis of microfluidic biochips. In *Design Automation Conference* (pp. 948–953).
86. Yao, H., Ho, T. Y., & Cai, Y. (2015). PACOR: Practical control-layer routing flow with length-matching constraint for flow-based microfluidic biochips. In *Proceedings of the Design Automation Conference* (pp. 142:1–142:6).
87. Yuh, P. H., Yang, C. L., & Chang, Y. W. (2007). BioRoute: A network-flow based routing algorithm for digital microfluidic biochips. In *International Conference on CAD* (pp. 752–757). Piscataway: IEEE Press.
88. Yuh, P. H., Yang, C. L., & Chang, Y. W. (2007). Placement of defect-tolerant digital microfluidic biochips using the T-tree formulation. *ACM JETC, 3*(3), 13.
89. Zheng, B., Roach, L. S., & Ismagilov, R. F. (2003). Screening of protein crystallization conditions on a microfluidic chip using nanoliter-size droplets. *Journal of the American Chemical Society, 125*(37), 11170–11171.

A New Ageing-Aware Approach via Path Isolation

Yue Lu, Shengyu Duan, and Tom J. Kazmierski

1 Introduction

With the drastic scaling of CMOS technology, design of robust system is becoming a major challenge. Among various reliability threats, Bias Temperature Instability (BTI) is emerging as one of the main reliability concerns in nano-scale technologies. Bias Temperature Instability (BTI) manifests itself as an increase in transistor threshold voltage (V_{th}) [1]. It degrades NMOS and PMOS transistors in the form of Positive BTI (PBTI) and Negative BTI (NBTI), respectively, but PBTI is only considered to be crucial for the CMOS technology beyond 45-nm due to the adoption of high-k materials [2]. For logic circuits, BTI-induced V_{th} shift results in path delay increase, eventually causing timing violations [3].

A great many techniques have been proposed to overcome BTI-induced timing errors at the design phase. Kumar et al. proposed to synthesize a circuit by using a re-characterized library that includes post-ageing information [4]. Wu and Marculescu exhaustively swap the input signals of a logic to find the minimum degradation circuit structure [5]. A gate-level approach is presented in [6], where a Lagrangian Relaxation algorithm is used to find the optimal size for each device, considering the BTI effect. These techniques may significantly improve circuit lifetime reliability, but require very high complexity. For this reason, they cannot be incorporated in a logic synthesis process. In practice, an over-design method is often used at the synthesis stage, typically by up-sizing the circuit to put in pessimistic timing

Y. Lu · S. Duan
School of Electronic and Computer Engineering, University of Southampton, Southampton, UK
e-mail: yl15g13@ecs.soton.ac.uk; yl15g13@ecs.soton.ac.uk

T. J. Kazmierski (✉)
University of Southampton, Southampton, UK
e-mail: tjk@ecs.soton.ac.uk

© Springer Nature Switzerland AG 2020
T. J. Kazmierski et al. (eds.), *Languages, Design Methods, and Tools for Electronic System Design*, Lecture Notes in Electrical Engineering 611,
https://doi.org/10.1007/978-3-030-31585-6_5

margins. The additional margin compensates for the delay increase caused by the BTI effect, improving the lifetime reliability. Although the over-design method is easy to implement, it usually induces larger-than-necessary overheads [7]. For sub-45-nm technologies, larger timing margins are needed due to the co-action of NBTI and PBTI, causing even greater costs. Therefore, it is necessary to develop alternative approaches for conventional over-design to ensure the circuit lifetime reliability with less overheads.

In this design, we propose an ageing-aware approach via path isolation to mitigate BTI degradation for given constraints. Specifically, we exploit all the suspicious paths that may lead to timing violations with the presence of ageing, and then insert the FFs (flip-flops) into these paths to realize isolation. With the guidance of our proposed synthesis algorithm, the area overhead can be significantly reduced by carefully gate sizing. This proposed approach is demonstrated on a 16-tap FIR filter design, simulation results show that, with the same given constraints, our design outperforms the conventional over-design techniques in terms of area saving.

The rest of the paper is organized as follows. Section 2 shows a brief review about ageing effects. Section 3 describes the principle of the proposed ageing aware approach. A case study of FIR filter and corresponding simulation results are analysed in Sect. 4 and Sect. 5 summarizes and concludes the paper.

2 BTI Effect and Delay Degradation

2.1 Transistor-Level BTI Modelling

The BTI effect is physically described as the consequence of charge generation on the transistor oxide interface [1]. The charges are produced when a transistor is turned on, and will be partially neutralized in the OFF state. These charges accumulate over time, resulting in a threshold voltage shift (ΔV_{th}). A simplified analytical model for BTI is presented in a previous paper [8], shown in Eq. 1.

$$\Delta V_{th}(t) = b.\alpha^n.t^n \tag{1}$$

where t is the operational time; b is a constant parameter determined by the technology node and the environmental conditions like supply voltage and temperature; α is the BTI stress duty cycle, given by the ratio of transistor ON time to the total; n is the time exponential constant, equal to 0.16, according to [9].

2.2 Circuit-Level BTI Degradations

For a logic gate, BTI-induced ΔV_{th} causes a linear increase to the signal propagation delay [10]. Thus, BTI-induced delay shift for a gate (D_{gate}) can be modelled by an equation, shown in Eq. 2, where K is dependent on technology node, supply voltage, temperature, *etc.*, and D_0 is the intrinsic gate delay.

$$\Delta D_{gate}(t) = K.D_0.\alpha^n.t^n \qquad (2)$$

A logic circuit is constructed by plenty of logic gates, each suffering from the BTI degradation. The overall degradation for a signal path of a circuit is therefore determined by the sum of delay shifts for all gates of the path. Based on Eq. 2, we give BTI-induced path degradation (ΔD_{path}) in Eq. 3, where $D_{0(i)}$ and α_i are the intrinsic delay and stress duty cycle for the gate i, respectively, and N indicates the total number of gates of the path.

$$\Delta D_{path}(t) = K.t^n.\sum_{i=1}^{N} D_{0(i)}.\alpha_i^n \qquad (3)$$

According to [11], the BTI effect causes 20% delay degradation after 10 years for a 65-nm circuit. This empirical data can be used to predict the delay increase after a specific operational time, by fitting the data into Eq. 3. While the stress duty cycle of each gate is highly dependent on the circuit workload, which may be hardly predicted at the design phase, accurate design-time ageing prediction is believed to be nearly impossible. One can only estimate the range of degradations based on the given conditions. According to [12], the difference of delay degradation is around 16% with different circuit workloads. Therefore, the minimum, typical and maximum percentage delay shifts for different operational times can be computed, shown in Table 1, for a 65-nm circuit.

In order to promise an expected lifetime considering different circuit workloads, a timing margin larger than the maximum delay degradation may be applied. For instance, a 25% guardband may be used to ensure a circuit working properly in 10 years. The pessimistic timing margin can be realized, causing great overhead if applying conventional over-design method. This motivates us to explore a less costly guardbanding method, as will be described in the following section.

Table 1 Minimum, typical and maximum percentage delay shifts due to BTI for 65-nm technology

Operational time (year)	Delay degradation (%)		
	Min.	Typ.	Max.
2	12.99	15.46	17.93
4	14.51	17.27	20.03
6	15.48	18.43	21.38
8	16.21	19.3	22.39
10	16.8	20	23.2

3 Proposed Ageing-Aware Approach

In this section, a novel approach is proposed to efficiently increase circuit robustness against ageing with negligible area and delay overheads. The basic idea is to isolate potential timing-violated paths with the presence of ageing into two cycles, thereby providing more guardband to mitigate ageing-issued timing violations. Depending on different design requirements about ageing, the circuit guardband can be adjusted ranging from 10% to 25%. Compared with standard two-cycle pipeline operation, the proposed approach inserts D-flip-flops as checkpoints into less number of circuit paths while acceptable timing margin is acquired to prevent ageing-issued problems.

3.1 Motivational Example

For the sake of simplicity, we choose a 8-bit ripple carry adder as an example, the path delay distribution across sum bits is showed in Fig. 1. The green bars represent the original path delay without ageing effects and the red bars denote the increased circuit delay due to circuit ageing, respectively. It can be clearly seen that the path delay of MSB violates timing constraints with the presence of ageing firstly. Once the MSB paths start to fail, it would lead to a rapid decline in the computation accuracy. With the adoption of proposed approach, this concern can be significantly eliminated. For a certain amount of guardband these violated paths would be separated into two parts to extend the circuit lifetime. In other words, the previous one-cycle operation is modified as two-cycle pipeline computation. In terms of performance loss, only one extra clock cycle is required without any effects on system sampling or throughput rate.

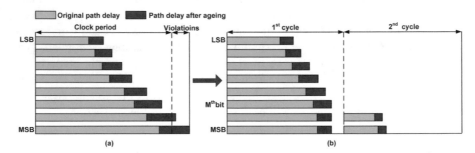

Fig. 1 Design time and post-ageing data arrival times before and after ageing-aware optimization for a 8-bit adder. (**a**) Path delay distribution after ageing. (**b**) Path delay distribution using critical path isolation

3.2 Algorithm for Path Isolation

In the proposed approach, our main objective is to assign ageing-aware timing constraints to the circuit via path isolation method during the synthesis stage to extend the circuit lifetime according to a certain amount of guardband. In order to realize the path isolations, a large number of flip-flops are inserted to cover all potential circuit paths. Taking the circuit's cost-efficiency into consideration, we propose a synthesis method via gate sizing to reduce the number of inserted flip-flops while maintaining the same path coverage. Specifically, for the circuit paths with the same end-point, the number of their shared path nodes grows with the increase of circuit depth while that of the total path node is reduced. Therefore, once a certain amount of guardband is given, we can move the path isolation nodes forward to the deeper circuit paths by upsizing and downsizing the different part of gates. Accordingly, the required insertion nodes can be significantly reduced, while the totalarea overhead is very small.

We propose Algorithm 1 to isolate some paths and put in a specific timing margin, compensating for the delay shift caused by the BTI effect. Specifically, the required intrinsic circuit delay D_{t0} is first computed based on the maximum delay allowed by the system and the percentage timing margin. In each iteration of the optimization, we focus on the most critical path, which has the largest delay and thus requires the greatest delay reduction. To isolate the path, the gates of the path are divided into a lower-depth group (G_l) and a greater-depth group (G_g) based on the signal arrival time on their output nodes: a gate with the output signal arrival time smaller/larger than D_{t0} is considered to be at the lower-/greater-depth and is categorized into G_l/G_g. One flip-flop (FF) is then inserted in the path to isolate the gates of G_l and G_g. In this way, the most critical path is broken down into two paths, both having the path delay smaller than D_{t0}. In order to ensure the correct function, other paths also need to be changed to support a two-cycle operation: FFs are inserted at the specific internal nodes of all paths that end at the same output as the most critical one, while one FF is required to be put in the each of the rest outputs, delaying the output signals for one clock cycle. We then use an algorithm to reduce the area overhead, as will be explained later. The above steps are repeated until the intrinsic circuit delay is not greater than D_{t0}, indicating that all paths have the required timing margin.

As has been described, the number of the required FFs may be reduced, by moving the FFs to a greater circuit depth. This can be realized by resizing the gates of a signal path. In specific, the FFs are inserted at the internal node of the critical path, where the signal arrival time is just smaller than D_{t0}, according to Algorithm 1. Therefore, by up-sizing some logic gates, the signal arrival times at the specific internal nodes may become smaller, moving the two-cycle operation point to a greater circuit depth. While the up-sizing process introduces more area on the combinational logic, it also potentially reduces the number of FFs required to isolate the paths, reducing the area of the sequential part (i.e. the FFs). Therefore, the overall circuit area needs to be evaluated to ensure a smaller area cost. We present

---

Algorithm 1 Path isolations for lifetime extension

Require: all paths have required timing margin;
 1: **procedure** PATHISO()
 2:     D_{t0} = max delay/(1+margin%);
 3:     $critPath$ = path with the largest delay;
 4:     **while** $D_{critPath} > D_{t0}$ **do**
 5:         **for each** $gate \in critPath$ **do**
 6:             **if** $arrival\ time < D_{t0}$ **then**
 7:                 $G_l = G_l \cup gate$;
 8:             **else**
 9:                 $G_g = G_g \cup gate$;
10:         Insert FFs based on G_l and G_g;
11:         AREAMIN();
12:         Identify new $critPath$;
13:     **return** $optCircuit$

Algorithm 2 Area cost minimization

 1: **procedure** AREAMIN()
 2:     Timing constraint = $D_{critPath}$;
 3:     **repeat**
 4:         Reduce timing constraint;
 5:         **upSize**(G_l) to meet timing constraint;
 6:         Re-insert FFs;
 7:         $Cost = AreaInc(\text{comb}) - AreaDec(\text{seq})$;
 8:     **until** $Cost > 0$;

Algorithm 2 to minimize the area cost. This algorithm iteratively tightens the timing constraint to up-size the gates of the critical path. Only the gates of G_l can be up-sized. This is because up-sizing the gates of G_g would not change the signal arrival time of any gates of G_l and thus, the FFs would be inserted at the same nodes as before the circuit is up-sized, causing more area cost. The FFs are re-inserted on the up-sized circuit according to the signal arrival time, similar to Algorithm 1. We compute the cost based on the area increase of the combinational logic and the area decrease of the sequential part. The optimization ends when the area cost becomes larger than 0, indicating the current design has the minimum area and any further change would induce more area cost.

Figure 2 shows a simple circuit to demonstrate the difference between the selections of D-flip-flop insertion nodes before and after gate sizing. It can be seen clearly, before circuit optimization, there are 4 two cycle D-flip-flop insertion nodes. After careful gate sizing the number of inserted registers is reduced twice while the guardband margin is maintained. In general, for the proposed ageing-aware synthesis approach, these constraints provide almost the same result as offered by conventional techniques, however, with a much less D-flip-flop insertion nodes.

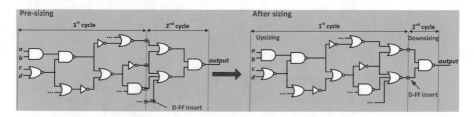

Fig. 2 An example of proposed synthesis approach

4 Case Study of a FIR Filter

To investigate the proposed ageing aware approach, we implemented a 16-tap digital filer datapath, which is an important application in hardware accelerations. With respect to different amount of guardband against ageing (5%–25%), the ageing-aware FIR circuit is modified respectively and the relevant simulation results are analysed in this section.

4.1 VLSI Implementation

For the sake of simplicity, we choose a FIR architecture with a 15% guardband as an example. Figure 3 shows a modified datapath of a 16-tap FIR filter with N-bit computation accuracy, where the red highlighted parts represent the auxiliary circuits used to realize path isolation. The optimal isolation nodes are found around the third adder stags, the particular amount of 15% guardband can be acquired by carefully gate sizing.

In this design, $2 \times M$ FFs are inserted to store the intermediate results, where the value of M is defined by the total number of timing violated paths affected by ageing. Since FIR filtering computation is based on multiplication and addition. Similar to the case of the 8-bit adder mentioned in Sect. 3, the critical paths are in the MSB of results and the path delay distribution for each individual end-point of all paths shows a degrade trend across result bits (from MSB to LSB groups). Therefore, in terms of an n-bit FIR filtering operation, fewer circuit paths would be seriously affected by ageing and lead to timing errors. If we assume all the circuit paths have 15% delay shift resulted from ageing problem. The results show that around $M + 2$-bit result would suffer from timing errors.

As seen in Fig. 3, 2*N+4-bit results are generated in an N-bit FIR filter operation to avoid overflows. The whole computation is implemented in a pipeline way. The 2*N+2-bit LSB group results are initially computed in the first clock cycle while some immediate results of MSB computation are stored to prevent timing errors. Until the next clock cycle, the complete results could be generated. It should be

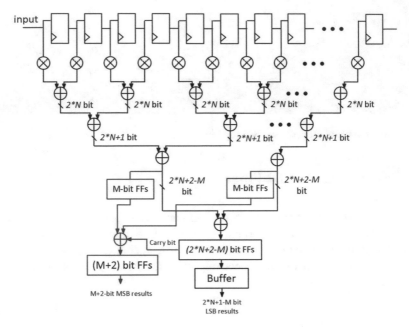

Fig. 3 An example of FIR datapath with 15% guardband

mentioned that during the second clock cycle, the MSB in 2*N+2-bit LSB group results would be sent to realize uncompleted operation as a carry bit.

4.2 Experimental Results

The proposed approach has been validated by applying it to a 16-tap digital FIR filter. We used Synopsys Design Compiler for logic-level optimization and our ageing-aware approach can provide guardband against ageing ranging from 5% to 25%.

Figure 4 shows the benefits of our proposed synthesis approach. As can be seen, the number of inserted FFs is significantly saved after circuit optimisation. In addition, these savings are kept rising with the increase of provided guardband. Compared with the conventional method, the number of FFs can be reduced by 17%.

As mentioned, practical approaches to improve circuit lifetime at the design stage is to leave enough timing margins against ageing effects, which increases the overheads of area and power. Thus, the circuit is over-designed. Here, we compare our proposed design with the ones that are over-designed for the same amount of guardband in terms of area overhead. The simulation results in Fig. 5 show that, with the increase of provided timing margin, our proposed design is more advantageous

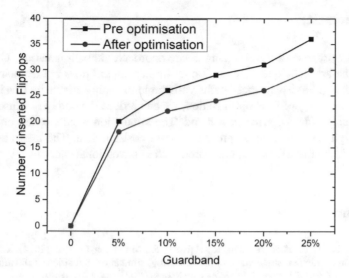

Fig. 4 Number of inserted flip-flop with increase of guardband for both before and after proposed synthesis approach

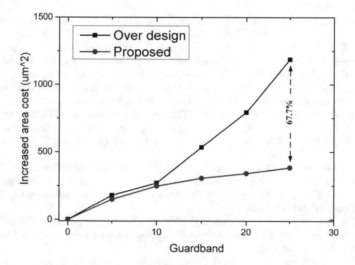

Fig. 5 Increased area cost with increase of guardband for both proposed and over designs

in area saving. As can be noted, our approach can save 35.3% area on average to guarantee the same lifetime reliability. When the timing guardband increase to 25%, the area saving can reach to 67.7%.

5   Conclusion

This paper proposes a novel ageing-aware approach via path isolation. Unlike the state-of-the-art techniques, the additional timing margin does not exclusively rely on gate sizing. We isolate these paths vulnerable to ageing effects by FFs insertion. Through careful gate-level optimisation, we can reduce the number of inserted FFs while maintaining the same guardband. The simulation results based on the FIR filter design show that our approach is more cost-efficient. The area cost of our approach is at most 67.7% less compared with a conventional over-design technique.

References

1. Sutaria, K., Ramkumar, A., Zhu, R., Rajveev, R., Ma, Y., & Cao, Y. (2014). BTI-induced aging under random stress waveforms: Modeling, simulation and silicon validation. In *51st ACM/EDAC/IEEE Design Automation Conference (DAC)* (pp. 1–6). IEEE.
2. Duhan, P., Rao, V. R., & Mohapatra, N. R. (2017). PBTI in HKMG nMOS transistors—effect of width, layout, and other technological parameters. *IEEE Transactions on Electron Devices, 64*(10), 4018–4024.
3. Fang, J., & Sapatnekar, S. S. (2013). The impact of BTI variations on timing in digital logic circuits. *IEEE Transactions on Device and Materials Reliability 13*(1), 277–286.
4. Kumar, S. V., Kim, C. H., & Sapatnekar, S. S. (2007). NBTI-aware synthesis of digital circuits. In *Proceedings of the 44th Annual Design Automation Conference* (pp 370–375). ACM.
5. Wu, K. C., & Marculescu, D. (2009). Joint logic restructuring and pin reordering against NBTI-induced performance degradation. In *Proceedings of the Conference on Design, Automation and Test in Europe* (pp. 75–80). European Design and Automation Association.
6. Paul, B. C., Kang, K., Kufluoglu, H., Alam, M. A., & Roy, K. (2007). Negative bias temperature instability: Estimation and design for improved reliability of nanoscale circuits. *IEEE Transactions on Computer-Aided Design of Integrated Circuits and Systems, 26*(4), 743–751.
7. Kang, K., Gangwal, S., Park, S. P., & Roy, K. (2008). NBTI induced performance degradation in logic and memory circuits: How effectively can we approach a reliability solution? In *Proceedings of the 2008 Asia and South Pacific Design Automation Conference* (pp. 726–731). IEEE Computer Society Press.
8. Wang, W., Wei, Z., Yang, S., & Cao, Y. (2007). An efficient method to identify critical gates under circuit aging. In *IEEE/ACM International Conference on Computer-Aided Design* (pp. 735–740). ICCAD 2007. IEEE.
9. Wu, K. C., & Marculescu, D. (2011). Aging-aware timing analysis and optimization considering path sensitization. In *Design, Automation & Test in Europe Conference & Exhibition (DATE)* (pp. 1–6). IEEE.
10. Chen, X., Wang, Y., Yang, H., Xie, Y., & Cao, Y. (2013). Assessment of circuit optimization techniques under NBTI. *IEEE Design & Test 30*(6), 40–49.
11. Ebrahimi, M., Oboril, F., Kiamehr, S., & Tahoori, M. B. (2013). Aging-aware logic synthesis. In *Proceedings of the International Conference on Computer-Aided Design* (pp. 61–68). IEEE Press.
12. Duan, S., Halak, B., & Zwolinski, M. (2017). An ageing-aware digital synthesis approach. In *14th International Conference on Synthesis, Modeling, Analysis and Simulation Methods and Applications to Circuit Design (SMACD)* (pp. 1–4). IEEE.

SystemC Coding Guideline for Faster Out-of-Order Parallel Discrete Event Simulation

Zhongqi Cheng, Tim Schmidt, and Rainer Dömer

1 Introduction

The IEEE SystemC standard [1] is widely used as a system level design language for specification, validation, and verification of complex system-on-chip models. With the rapidly growing complexity of embedded systems, a faster simulation of SystemC models is of high demand to shorten the design cycle.

The official proof-of-concept Accellera SystemC simulator [2] is based on Discrete Event Simulation (DES), which executes the SystemC model sequentially. This means that only one thread is allowed to run at any time during the simulation. Consequently, when running the Accellera SystemC simulator on a modern multi- or many-core processor, all but one cores remain idle and the parallel computation capabilities are largely wasted.

Parallel Discrete Event Simulation (PDES) [3] has gained significant attention because it can exploit the parallel computation power of modern processors and provide faster simulation. However, regular PDES is synchronous. Earlier completed simulation threads need to wait until all the other threads have reached the same simulation cycle barrier to continue their simulation. This strict total order still imposes a limitation on high performance parallel simulation.

Out-of-Order Parallel Discrete Event Simulation (OoO PDES) [4] was proposed for a better utilization of the parallel computation power. In OoO PDES, the simulation time is local to each thread, and thus, the global simulation cycle barrier is removed. Independent threads can execute in parallel even if they are in different time cycles.

Z. Cheng (✉) · T. Schmidt · R. Dömer
Center for Embedded and Cyber-Physical Systems, University of California, Irvine, CA, USA
e-mail: zhongqc@uci.edu; schmidtt@uci.edu; doemer@uci.edu

© Springer Nature Switzerland AG 2020
T. J. Kazmierski et al. (eds.), *Languages, Design Methods, and Tools for Electronic System Design*, Lecture Notes in Electrical Engineering 611,
https://doi.org/10.1007/978-3-030-31585-6_6

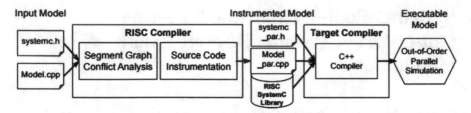

Fig. 1 RISC compiler and simulator for OoO PDES of SystemC [5]

The Recoding Infrastructure for SystemC (RISC) [5] provides a dedicated SystemC compiler and an advanced OoO PDES simulator for SystemC. RISC is available as an open-source project and can be downloaded freely from the official website [6]. Figure 1 shows the tool flow of RISC. The RISC compiler is used as a frontend to process the input SystemC file. It statically analyzes and derives a Segment Graph (SG) representation of the model. Based on the SG, the compiler is able to analyze data conflicts and event notifications among segments, and it instruments the information as multiple lookup tables into an intermediate model. This model is then linked against the OoO PDES library to generate an executable. During the simulation, every thread executes a sequence of segments along a path over the segment graph. The simulator dynamically checks the instrumented tables to make correct thread dispatching decisions, preserving the simulation semantics and timing accuracy.

1.1 Related Work

Various approaches have been proposed to further improve the simulation speed of OoO PDES. A segment-aware thread dispatching algorithm is studied in [7]. It takes into account the execution time for a specific segment as a prediction of the next run time, so that the dispatcher more accurately predicts the run time of the thread segments ahead and makes better dispatching decisions.

In [8], the authors extended the RISC compiler with the Port Call Path (PCP) technique, which reduces false positive conflicts in the channel analysis and significantly increases the simulation speed.

PDES was also studied in [9]. The authors proposed a conservative synchronous parallel simulation approach and a SystemC framework to speed up tightly coupled MPSoC simulations on multi-core hosts.

In [10], the authors proposed an open-source framework called systemc-clang for analyzing SystemC models with a mixture of register-transfer level and transaction-level components.

In this paper, we propose a coding guideline for SystemC users to build models with higher parallel potential that can be executed faster by the OoO PDES simulator. Specifically, the guideline suggests for users to insert extra **wait** statements into the model, so as to increase the granularity of the SG. With the finer granularity SG, variable and event conflicts can be constrained into shorter segments, thereby reducing the time of sequential execution, which is necessary, for example, during communication between modules in the system.

Our contributions in this work are summarized as follows:

1. We propose a formal metric ψ to estimate the level of parallelism of the model under OoO PDES.
2. We propose a coding guideline for the SystemC model designers to optimize the model for faster simulation.
3. We demonstrate that the proposed coding guideline enables significant speedup of OoO PDES.

2 SG Granularity and Simulation Speed

In OoO PDES, models are simulated at segment level. SG is described in details in Sect. 3.1. As shown in Fig. 2, module M has two sc_threads th1 and th2, and a member variable a. f() and g() are data crunching functions which work on local variables. The corresponding SG is shown in Fig. 3. Due to the data hazard over a, the two segments are not allowed to run in parallel. Figure 4 shows the scheduling of execution of the two sc_threads.

By inserting two new **wait** statements into the sc_threads, as shown in Fig. 5, the SG becomes Fig. 6. In this model, functions f() and g() are no longer in the same segment of the statements that access the shared variable a. Because f() and g() are conflict-free, they can now be executed in parallel as shown in Fig. 7, which significantly speeds up the simulation.

Fig. 2 Coarse grained source code

```
1   SC_MODEULE (M) {
2       ...
3       int a;
4       void th1 () {
5           a=1;
6           f ();
7       }
8
9       void th2 () {
10          g ();
11          a=2;
12      }
13      ...
14  }
```

Fig. 3 SG of Fig. 2

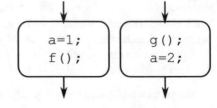

Fig. 4 Scheduling of Fig. 2

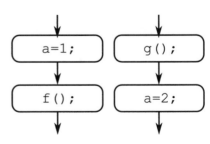

Fig. 5 Fine grained source code

```
1   SC_MODEULE(M) {
2       ...
3       int a;
4       void th1(){
5           a=1;
6           wait(
            SC_ZERO_TIME);
7           f();
8       }
9
10      void th2(){
11          g();
12          wait(
            SC_ZERO_TIME);
13          a=2;
14      }
15      ...
16  }
```

Fig. 6 SG of Fig. 5

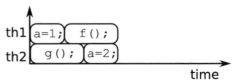

Fig. 7 Scheduling of Fig. 5

This leads to the conclusion that by increasing the granularity of SG, more code statements can run in parallel, and consequently increase the level of parallelism of a model and further speed up the simulation.

In the following section, we will show more details to confirm this idea and propose a coding guideline for the model designer to increase the parallel potential of the SystemC models under OoO PDES.

3   Recoding Infrastructure for SystemC

The fundamentals about RISC [6] are reviewed in this section for a better comprehension of the proposed coding guideline.

3.1   Segment Graph

The SG is the foundation for both static analysis by the RISC compiler and OoO PDES by the RISC simulator. It is built on top of the Abstract Syntax Tree (AST) of the input SystemC model.

A SG is a directed graph. Each node is called a segment, which represents the code statements executed during the simulation between two scheduling steps, i.e., the entry into the simulator kernel due to a **wait** statement in SystemC. The edges in SG represent the transition between segments. An example of SystemC source code and corresponding SG is shown in Figs. 8 and 9.

Fig. 8 Example source code

```
1   SC_MODEULE (M) {
2        index++;
3        wait(2,SC_NS);
4        k=1;
5        if(flag){
6            x++;
7            wait(
                 10,SC_NS);
8            y++;
9        }else{
10           a=5;
11       }
12       s=s*s;
13       wait(1,SC_NS);
14       t=s+1;
15   }
```

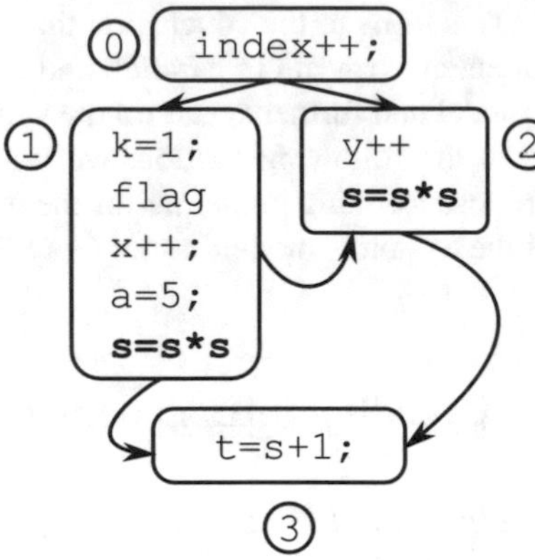

Fig. 9 SG of Fig. 8

Fig. 10 Data conflict table
for Fig. 8

In this example, line 8 y++ and line 12 s=s*s could be possibly executed in the same simulation cycle by a thread, so they are put both into segment2. One statement may also belong to multiple segments as it may occur on different simulation cycles. Both segment 1 and 2 have s=s*s in the above example. Note that a new segment starts only on **wait** statements except for the first one. The first segment is the entry point of a thread.

3.2   Data and Event Conflicts

The data conflict analysis takes place after the construction of the SG. It is automatically performed by the RISC compiler. Data conflicts between segments are caused by data hazards, i.e., parallel or out-of-order accesses to shared variables. There are three types of data hazards: read-after-write (RAW), write-after-write (WAW), and write-after-read (WAR). In the example in Fig. 3, segment 1 and 2 have data conflict due to the data hazard over the variable s. The RISC compiler checks the data conflicts between every pair of segments, and stores the result in a Data Conflict Table (DCT). Figure 10 shows the DCT for the example in Fig. 8. The red box indicates a conflict, and the blank ones mean conflict-free.

During the simulation, the OoO PDES simulator looks up the data conflict table to make safe thread dispatching decisions. If the segments of two ready-to-run threads have data conflicts, the thread with an earlier timestamp is dispatched by the scheduler. In general, segments with data conflicts are not allowed to execute in parallel.

Event and timing conflicts are two other kinds of conflicts that are taken care of in OoO PDES. They are analyzed in a similar fashion as the data conflict. Details are described in depth in [4], but omitted here for brevity.

4 Proposed Coding Guideline

In this section, we propose a new coding guideline for the SystemC model designers to write SystemC models with higher parallel simulation potential. Before describing the guideline, we first define a metric to estimate the level of parallelism of a SystemC model under OoO PDES.

4.1 Estimation for Level of Parallelism

The level of parallelism ψ is estimated as the amount of code statement pairs that can potentially execute in parallel. In OoO PDES, only code statements that belong to conflict-free segments can run in parallel, and hence, our estimation is expressed as:

$$\psi = \sum_i \sum_{\substack{j>i \\ \text{th}_i \neq \text{th}_j}} \text{HASNOCONFLICT}(\text{seg}_i, \text{seg}_j) \tag{1}$$

where i and j are the index of code statements in the model. seg_n is the segment that includes the nth code statement. And similarly, th_i is the thread that executes the nth code statement. Each single thread executes sequentially, and code statement i and j cannot execute in parallel if they belong to the same thread. $\text{HASNOCONFLICT}(\text{seg}_i, \text{seg}_j)$ returns 1 if seg_i and seg_j are conflict-free; otherwise, it returns 0.

If two segments are in conflict, then any pair of code statements that belong to the two segments are not allowed to execute in parallel, which would reduce ψ. Thus, the larger ψ is, the higher is the parallelism level of the input model.

4.2 Motivation

Our idea is motivated by the following observation:

Consider we have two segments: seg_1 and seg_2, which are executed by two different threads. There are, respectively, p and q statements in seg_1 and seg_2. ψ for this model is simply $\psi_1 = p \times q \times \text{HASNOCONFLICT}(seg_1, seg_2)$.

Now, if a **wait** statement is inserted into seg_1, such that seg_1 is partitioned into two non-overlapping segments: seg_{11} and seg_{12}. After the partitioning, seg_{11} includes the first p_1 statements of seg_1, and seg_{12} includes the other $p_2 = p - p_1$ statements of seg_1. ψ for the new model becomes $\psi_2 = p_1 \times q \times \text{HASNOCONFLICT}(seg_{11}, seg_2) + p_2 \times q \times \text{HASNOCONFLICT}(seg_{12}, seg_2)$. seg_{11} and seg_{12} are executed by the same thread, and hence, they must run sequentially and ψ_2 does not increase.

When comparing ψ_1 and ψ_2, we get four different scenarios:

1. The conflict between seg_1 and seg_2 is only incurred by certain state-ments in the first p_1 statements of seg_1, and the last p_2 statements are conflict-free. This indicates that $\text{HASNOCONFLICT}(seg_{11}, seg_2) = 0$, $\text{HASNOCONFLICT}(seg_{12}, seg_2) = 1$, and $\text{HASNOCONFLICT}(seg_1, seg_2) = 0$. Under this scenario, $\psi_1 = 0$ and $\psi_2 = p_2 \times q$. ψ_2 is larger than ψ_1.
2. The conflict between seg_1 and seg_2 is only incurred by certain statements in the last p_2 statements of seg_1, and the other p_1 statements are conflict-free. This indicates that $\text{HASNOCONFLICT}(seg_{11}, seg_2) = 1$, $\text{HASNOCONFLICT}(seg_{12}, seg_2) = 0$, and $\text{HASNOCONFLICT}(seg_1, seg_2) = 0$. Under this scenario, $\psi_1 = 0$ and $\psi_2 = p_1 \times q$. ψ_2 is larger than ψ_1.
3. The conflict between seg_1 and seg_2 is incurred both by certain statements in the first p_1 statements and the other p_2 statements of seg_1. This indicates that $\text{HASNOCONFLICT}(seg_{11}, seg_2) = 0$, $\text{HASNOCONFLICT}(seg_{12}, seg_2) = 0$, and $\text{HASNOCONFLICT}(seg_1, seg_2) = 0$. Under this scenario, $\psi_1 = 0$ and $\psi_2 = 0$. ψ_2 is equal to ψ_1.
4. seg_1 and seg_2 are conflict-free. This indicates that $\text{HASNOCONFLICT}(seg_{11}, seg_2) = 1$, $\text{HASNOCONFLICT}(seg_{12}, seg_2) = 1$, and $\text{HASNOCONFLICT}(seg_1, seg_2) = 1$. Under this scenario, $\psi_1 = p \times q$ and $\psi_2 = p_1 \times q + p_2 \times q = p \times q$. ψ_2 is equal to ψ_1.

The four scenarios suggest that

1. Partitioning a segment does not decrease the parallel potential of a model.
2. If the user carefully selects the place to insert the extra segment boundary, i.e., **wait** statement, ψ can be increased significantly and results in a model with higher parallelism level.

4.3   Overhead Consideration

One may deduce that it is always beneficial to insert as many extra **wait** statements as possible, because by doing this the ψ of the model keeps increasing. Although the deduction is correct, it is not a good practice.

Each extra **wait** statement will increase the number of segments in the segment graph by one. And the size of conflict tables is to the square of the segment count. Thus, if too many extra **wait** statements are inserted, the time cost for static analysis and dynamic checking will grow dramatically, which would rather decrease the simulation performance. Besides, too many extra **wait** statements may also make the model incomprehensible.

Last but not least, each new **wait** statement creates an extra scheduler entry point into the simulator kernel which incurs high overhead.

4.4 Suggestions

Motivated by the above observations and considerations, we propose the following suggestions for the SystemC model designers to properly place extra **wait** statements in the source code, so as to increase the parallel potential of the model under OoO PDES.

4.4.1 Use the Wait-for-Delta-Cycle Primitive as the Extra Segment Boundary

There are six different kinds of **wait** primitives in the SystemC standard [1]:

1. **wait**() : Wait for the sensitivity list event to occur.
2. **wait**(int) : Wait for n clock cycles in SC_CTHREAD.
3. **wait**(event) : Wait for the event mentioned as parameter to occur.
4. **wait**(double,sc_time_unit) : Wait for specified time.
5. **wait**(double,sc_time_unit, event) : Wait for specified time or event to occur.
6. **wait**(SC_ZERO_TIME): Wait for one delta cycle.

The event related **wait** primitives shall not be used because they require proper events to be notified. For the wait-for-time primitive, it is likely to change the simulation time cycle, which is not desirable. Thus, in order to maintain the semantics and timing accuracy of the original SystemC model, we suggest to the designers to use wait-for-delta-cycle primitive, i.e., **wait**(SC_ZERO_TIME) as extra segment boundaries.[1]

4.4.2 Partition the Heavy Segments

As mentioned in Sect. 4.3, the cost for one extra **wait** statement is independent of where it is inserted. Thus, in order to maximize the gain of ψ of the model, we

---

[1]Note that the timing accuracy of a robust model will not be affected by extra delta cycles.

suggest the users to partition computational intensive segments, which we refer to as *heavy segments*.

Unfortunately, it is not obvious to identify heavy segments directly from the model code. However, the RISC compiler is able to dump the statically generated SG and the DCT into files by turning on the `-risc:dump` command line option. The SG is then dumped into a `.dot` file which can be viewed graphically using the `xdot.py` tool. Also, the DCT is dumped into an HTML file which the designer can easily view in any browser. An example SystemC source code is shown in Fig. 11. The dumped SG and DCT are shown in Figs. 12 and 13. The level of parallelism ψ for this model is $\psi 1 = 6 + 5 = 11$.

Fig. 11 Source code for module M

```
1   SC_MODEULE(M)
2   {
3       int c;
4       void th1()
5       {
6           int x=1;
7           wait(10,SC_NS);
8           c=42;
9
10          for(int i=0;
                i<100;
11              i++) x++;
12      }
13      void th2()
14      {
15          int y=100;
16          wait(1,SC_NS);
17          c=0;
18
19          for(int j=0;
                j<100;
20              j++) y--;
21      }
22      ...
23  }
```

Fig. 12 SG for Fig. 11

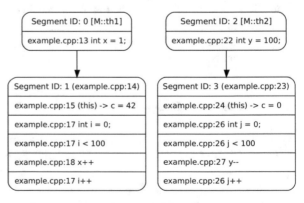

Fig. 13 DCT for Fig. 11

(Seg,Inst)	(0,0)	(1,0)	(2,0)	(3,0)
(0,0)				
(1,0)		M::c, 0		M::c, 0
(2,0)				
(3,0)		M::c, 0		M::c, 0

Fig. 14 Source code for module M after partitioning

```
1   SC_MODEULE(M)
2   {
3       int c;
4       void th1()
5       {
6           int x=1;
7           wait(10,SC_NS);
8           c=42;
9           wait(SC_ZERO_TIME);
10          for(int i=0;
                i<100;
11              i++) x++;
12      }
13      void th2()
14      {
15          int y=100;
16          wait(1,SC_NS);
17          c=0;
18          wait(SC_ZERO_TIME);
19          for(int j=0;
                j<100;
20              j++) y--;
21      }
22      ...
23  }
```

From the SG, it is apparent that segment 1 and segment 3 are heavy segments which both contain loops. In order to increase the parallelism level of the model, we wish to partition the conflict-free statements from the conflicting ones in the segment, as described in the first and second scenarios in the previous section. To locate the conflicting statement, the user can refer to the dumped Data Conflict Table. In the ((1,0),(3,0)) entry of the table, it shows that the data conflict is over the variable **M::c**, and so the conflict is between statement lines 8 and 17 in Fig. 11.[2] In this example, the conflicting statements are not inside the computationally intensive code pieces, that are, the for loops. So we can partition the segments by inserting **wait** statements after lines 9 and 18. The optimized model is shown in Fig. 14. The dumped SG and DCT are shown in Figs. 15 and 16. Now, the level of parallelism ψ becomes $\psi 2 = 6 + 5 + 4 \times 6 = 35$. The parallel potential is further intensified during the simulation due to the two conflict-free **for** loops.

---

[2]The instance id is shown here, which is not of interest in this paper.

Fig. 15 SG for Fig. 14

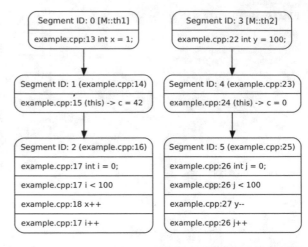

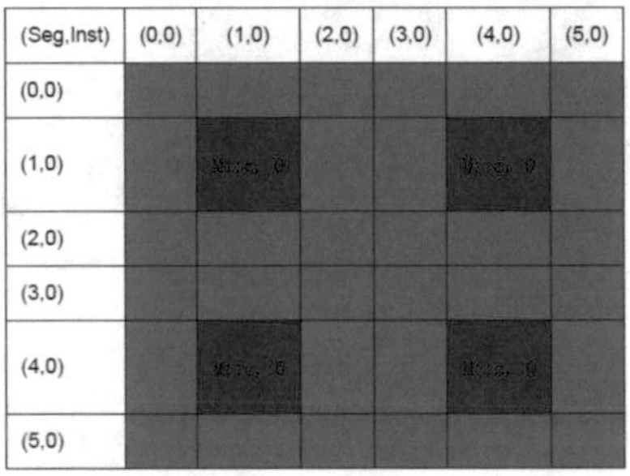

Fig. 16 DCT for Fig. 14

5   Experiments and Results

We have applied the proposed coding guideline to several SystemC model examples. We first tested it on the synthetic benchmarks generated by the TGFF tool to validate the effectiveness of our coding guideline. Then, we evaluate the guideline with two real world designs, Canny Edge Detector and Audio/Video Decoder, to demonstrate the performance. The experiments are performed on an Intel E3-1240 host machine, which has a total of 8 cores (4 cores with 2-way hyperthreading each). The CPU frequency scaling is turned off so as to obtain repeatable results.

5.1   TGFF Benchmarks

We first examine the performance of the proposed coding guideline on a synthetic benchmark, which is automatically generated by the TGFF tool with SystemC extension [7]. Figure 17 shows the data flow block diagram of the generated model. It has a source and a sink, and multiple parallel lanes of nodes in between. Figure 18 shows the source code for each node. Each node module first gets an input from

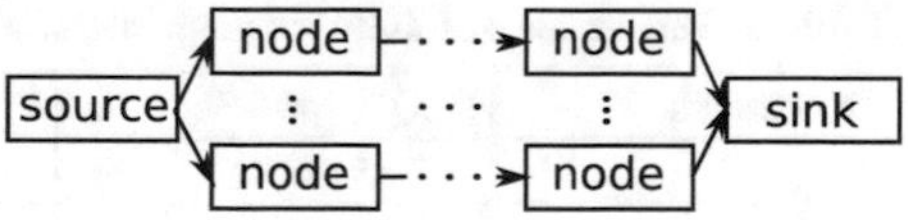

Fig. 17 Block diagram of TGFF models

Fig. 18 Original source code of generated Testbench model

```
1  SC_MODEULE(Node)
2  {
3      sc_port<input> in;
4      sc_port<output> out;
5      void th1()
6      {
7          int a = in.read();
8          for(int i=0;
9              i<WORKLOAD;
10             i++) a++;
11         out.write(a);
12     }
13     ...
14 }
```

Fig. 19 Optimized source code of generated Testbench model

```
1  SC_MODEULE(Node)
2  {
3      sc_port<input> in;
4      sc_port<output> out;
5      void th1()
6      {
7          int a = in.read();
8          wait(
           SC_ZERO_TIME);
9          for(int i=0;
10             i<WORKLOAD;
11             i++) a++;
12         wait(
           SC_ZERO_TIME);
13         out.write(a);
14     }
15     ...
16 }
```

a channel, and then does data crunching which is computationally intensive. The data crunching accesses only local variables and thus is conflict-free. After the computation the module outputs the result to another channel. In such model, data conflicts are incurred only by channel communications, which are caused by the parallel accesses to the shared variables in the channels. To optimize the model, we apply the proposed coding guideline and put **wait**(SC_ZERO_TIME) statements around the data crunching parts. The source code for the optimized module is shown in Fig. 19.

Through a parameter to the TGFF generator, we are able to control the total number of lanes as well as nodes per lane, and each lane may consist of various number of nodes. The data crunching workload of each node is controlled by the number of iterations of the **for** loop.

Table 1 Performance of TGFF benchmarks, simulator run times [sec], and CPU utilization

Benchmark	SEQ	PAR	GDL
1	63.55 (99%)	17.85 (377%)	10.48 (690%)
2	63.54 (99%)	17.63 (379%)	10.91 (663%)
3	134.41 (99%)	88.41 (155%)	81.55 (172%)
4	349.86 (99%)	165.41 (214%)	93.44 (400%)
5	493.02 (99%)	169.12 (301%)	99.17 (552%)
6	134.40 (99%)	92.00 (155%)	81.10 (173%)
Average	206.46 (99%)	91.74 (263.5%)	62.77 (441%)

We studied 6 test cases with different data flow configurations in this experiment. Table 1 shows the performance of the simulations before and after applying the coding guideline. The first column SEQ refers to the sequential simulation with the reference Accellera SystemC simulator. Under the sequential simulation, the CPU utilization is always below 100% because only one thread is running at any time during the simulation. The second column PAR refers to the OoO PDES before applying the coding guideline. It shows that on average, the simulation of the original models is 2.3x faster than SEQ. The third column GDL refers to the OoO PDES after applying the coding guideline. It is 3.2x faster than SEQ, and 1.4x faster than PAR. For the first benchmark, GDL achieved a maximum speedup of 1.7x over PAR, and the latter one is 3.5x faster than SEQ. Note that the CPU utilization is larger than the speedup over SEQ. This is because in OoO PDES there is some overhead for checking conflict tables. The results confirm that our coding guideline can be very effective in achieving higher speedup under OoO PDES.

5.2 Real World Examples

We then evaluate the proposed coding guideline with two real world examples, namely Canny Edge Detector and Audio/Video Decoder modeled similarly to the benchmarks used in [7] and [8].

5.2.1 Canny Edge Detector

Our first real world example is the Canny edge detector, which filters edges in an image. The edge detector is a structurally five-stage pipeline, and each stage has a communication-computation-communication code structure. Communication between two pipeline stages is via a user-defined channel in which the read and write functions access the shared channel variable. In this experiment, a sequence of 20 images is fed into the pipeline and correspondingly generates 20 outputs. The outputs are verified to ensure a correct simulation.

Table 2 Performance of Canny edge detector

	SEQ	PAR	GDL
Simulation time (s)	24.85	19.96	17.23
CPU utilization	100%	127%	149%
Speedup	1.00	1.24	1.44

Fig. 20 Block diagram of Audio/Video decoder

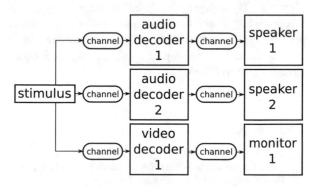

Table 3 Performance of Audio/Video decoder

	SEQ	PAR	GDL
Simulation time (s)	73.41	48.24	26.67
CPU utilization	100%	152%	247%
Speedup	1.00	1.52	2.75

Table 2 shows the simulation time and CPU utilization before and after applying the coding guideline. By using the original model, a CPU utilization of 127% is achieved, which is due to the conflicts among communications. With the optimized model, the CPU utilization is increased to 149%, and the OoO PDES speed is increased by 1.2x. The speedup is not as impressive as in the TGFF test cases. This is because the workload of each pipeline stage varies greatly, and the bottleneck of the simulation speed is determined by the longest stage. However, this experiment still confirms the effectiveness of the proposed coding guideline.

5.3   A/V Decoder

The second real world test case is an Audio/Video decoder. The model structure is shown in Fig. 20. The stimulus sends the encoded stream to one video decoder and the left and right audio decoders. Then, the video decoder outputs the result to a monitor, and the audio decoders output the results to two speakers. The results for this test case are shown in Table 3. The execution times cost for OoO PDES before and after applying the coding guideline are 48.24 s and 26.67 s, which suggest the optimized model executes 1.8x faster. The speedup is reasonable because the encoding and decoding stages have similar computation loads. The result again confirms the effectiveness of the proposed coding guideline.

6 Conclusion

In this paper, we proposed a coding guideline for the SystemC model designers who use OoO PDES parallel execution enabled by the Recoding Infrastructure for SystemC. By applying the coding guideline, the granularity of the Segment Graph becomes larger, and thus results in a faster execution speed. Our experiments show that by applying the proposed coding guideline, the optimized SystemC model is able to achieve a speedup of up to 1.7x on an 8 core machine, on top of the 3.5x speedup due to PDES.

References

1. IEEE Standard 1666–2011 for Standard SystemC® Language Reference Manual. (2012). IEEE Computer Society, January 2012.
2. SystemC Language Working Group. (2014). SystemC 2.3.1, Core SystemC Language and Examples, Accellera Systems Initiative. [Online]. Available: http://accellera.org/downloads/standards/systemc, 2014.
3. Fujimoto, R. (1990). Parallel discrete event simulation. *Commun. ACM, 33,* 3053.
4. Chen, W., Han, X., Chang, C. W., Liu, G., & Dömer, R. (2014). Out-of-order parallel discrete event simulation for transaction level models. *IEEE Transactions on Computer-Aided Design of Integrated Circuits and Systems, 33*(12), 1859–1872.
5. Dömer, R., Liu, G., & Schmidt, T. (2016). Parallel simulation. In S. Ha, & J. Teich (Eds.), *Handbook of Hardware/Software Codesign.* Dordrecht: Springer.
6. Lab for Embedded Computer Systems (LECS). Recoding Infrastructure for SystemC [Online]. Available: www.cecs.uci.edu/~doemer/risc.html#RISC050.
7. Liu, G., Schmidt, T., & Dömer, R. (2016). A segment-aware multi-core scheduler for systemC PDES. In *Proceedings of the International High Level Design Validation and Test Workshop,* Santa Cruz, California, October 2016.
8. Schmidt, T., Cheng, Z., & Dömer, R. (2018). Port call path sensitive conflict analysis for instance-aware parallel systemc simulation. In *Proceedings of Design, Automation and Test in Europe,* Dresden, Germany, March 2018.
9. Schumacher, C., Leupers, R., Petras, D., & Hoffmann, A. (2010). parSC: synchronous parallel systemc simulation on multi-core host architectures. In *Proceedings of the International Conference on Hardware/Software Codesign and System Synthesis* (pp. 241–246).
10. Kaushik, A., & Patel, H. D. (2013). SystemC-clang: an open-source framework for analyzing mixed-abstraction SystemC models. In *Proceedings of the Forum on Specification and Design Languages (FDL),* Paris, 2013.

Extensible and Configurable RISC-V Based Virtual Prototype

Vladimir Herdt, Daniel Große, Hoang M. Le, and Rolf Drechsler

1 Introduction

Enormous innovations are enabled by the *Internet-of-Things* (IoT) since every device is connected to the Internet. Forecasts see additional economic impact resulting from Industrial IoT. In the last years the complexity of IoT devices has been increasing steadily with various conflicting requirements. On the one hand, IoT devices need to provide smart functions with a high performance including real-time computing capabilities, connectivity, and remote access as well as safety, security, and high reliability. At the same time they have to be cheap, work efficiently with an extremely small amount of memory and limited resources, and should further consume only a minimal amount of power to ensure a very long lifetime.

To meet the requirements of a specific IoT system, a crucial component is the processor. Stimulated from the enormous momentum of open source software, a counterpart on the hardware side recently emerged: *RISC-V* [17, 18]. RISC-V is an open source *Instruction Set Architecture* (ISA) which is license-free and royalty-free. The ISA standard is maintained by the non-profit RISC-V foundation and is appropriate for all levels of computing systems, i.e., from micro-controllers to supercomputers. The RISC-V ecosystem is rapidly growing, ranging from HW, e.g., various HW implementations (free as well as commercial) to high-speed *Instruction Set Simulators* (ISSs). These ISSs facilitate functional verification of

V. Herdt (✉) · H. M. Le
Institute of Computer Science, University of Bremen, Bremen, Germany
e-mail: vherdt@informatik.uni-bremen.de; hle@informatik.uni-bremen.de

D. Große · R. Drechsler
Institute of Computer Science, University of Bremen and Cyber-Physical Systems, DFKI GmbH, Bremen, Germany
e-mail: grosse@informatik.uni-bremen.de; drechsle@informatik.uni-bremen.de

© Springer Nature Switzerland AG 2020
T. J. Kazmierski et al. (eds.), *Languages, Design Methods, and Tools for Electronic System Design*, Lecture Notes in Electrical Engineering 611,
https://doi.org/10.1007/978-3-030-31585-6_7

RTL implementations as well as early SW development to some extent. However, being designed predominantly for speed, they can hardly be extended to support further system-level use cases such as design space exploration, power/timing/performance validation, or analysis of complex HW/SW interactions.

A major industry-proven approach to deal with these use cases in earlier phases of the design flow is to employ *Virtual Prototypes* (VPs) [10] at the abstraction of *Electronic System Level* (ESL) [1]. In industrial practice, the standardized C++-based modeling language SystemC and *Transaction Level Modeling* (TLM) techniques [4, 9] are being heavily used together to create VPs. Depending on the specific use case, advanced state-of-the-art SystemC-based techniques beyond functional modeling (see e.g. [5–7, 11, 16]) are to be applied on top of the basic VPs. The much earlier availability as well as the significantly faster simulation speed in comparison to RTL are among the main benefits of SystemC-based VPs.

In this paper, we propose and implement the first RISC-V based VP to further expand and bring the benefits of VPs to the RISC-V ecosystem. With the goal of filling the mentioned gap in supporting further system-level use cases, SystemC is necessarily the language of choice. The VP is therefore implemented in standard-compliant SystemC and TLM-2.0 and designed as extensible and configurable platform with a generic bus system. We provide a RISC-V RV32IM core and a PLIC-based interrupt controller with an essential set of peripherals. We demonstrate the extensibility of our VP by two examples: addition of a sensor peripheral and extension by GDB debug functionality from the application SW perspective. In the experimental evaluation we show the high simulation performance of our VP based on several optimizations. Our RISC-V VP is fully open source[1] to stimulate further research and development.

Related Work

As mentioned earlier, the RISC-V ecosystem already has various high-speed ISSs such as the reference simulator Spike [15], RISCV-QEMU [12], or RV8 [13]. They are mainly designed to simulate as fast as possible and predominantly employ dynamic binary translation (to x86_64) techniques. This is however a trade-off as accurately modeling power or timing information for instructions becomes much more challenging. The full-system simulator gem5 [2], at the time of writing also has initial support for RISC-V. gem5 provides more detailed models of processors and memories and can in principle also be extended for accurate modeling of extra-functional properties. However, it employs a different modeling style and thus hinders the integration of advanced SystemC-based techniques. The project SoCRocket [14] that develops an open source SystemC-based VP for the SPARC

---

[1] Available at https://github.com/agra-uni-bremen/riscv-vp, for more information and updates also visit www.systemc-verification.org/riscv-vp.

V8 architecture can be considered comparable to our effort. Finally, commercial VP tools such as Synopsys Virtualizer or Mentor Vista might also support RISC-V but their implementation is proprietary.

2 Preliminaries

2.1 RISC-V

RISC-V is an open and free instruction set architecture (ISA). The ISA consists of a mandatory base integer instruction set and various optional extensions. The integer set is available in three different configurations with 32, 64, and 128 bit width registers, respectively: RV32I, RV64I, and RV128I. Additionally, the RV32E configuration, which is essentially a lightweight RV32I with a reduced number of registers, is available and intended for (very) small embedded devices. Extensions are denoted with a single letter, e.g., M (integer multiplication and division), A (atomic instructions), C (compressed instructions), etc. A comprehensive description of the RISC-V instruction set is available in the specification [17].

The second volume of the RISC-V ISA specification defines a privileged architecture description [18]. It defines *control and status registers* (CSRs), which are registers serving a special purpose. For example the *misa* (Machine ISA) register is a read-only CSR that contains information about the supported ISA. Another example is the *mtvec* (Machine Trap-Vector Base-Address) CSR that stores the address of the trap/interrupt handler. The privileged architecture description provides a small set of instructions for interrupt handling (*wfi*, *mret*) and interacting with the system environment (*ecall*, *ebreak*).

2.2 SystemC and TLM

SystemC is a C++ class library that includes an event-driven simulation kernel. The structure of a SystemC design is described with ports and modules, whereas the behavior is described in processes which are triggered by events. The execution of a process is non-preemptive, i.e., the kernel receives the control back if the process has finished its execution or suspends itself by calling *wait()*. SystemC provides three types of processes with SC_THREAD being the most general type, i.e., the other two can be modeled by using SC_THREAD. For event-based synchronization, SystemC offers many variants of *wait()* and *notify()* such as *wait(time)*, *wait(event)*, *event.notify(delay)*, *event.notify()*, etc.

Communication between modules is implemented through (TLM) transactions. A transaction object essentially consists of the command (e.g., read/write), the start address, the data length, and the data pointer. It allows to implement various memory access operations. Optionally, a transaction can be associated with a delay (modeled as *sc_time* data structure), which denotes the execution time of the transaction and allows to obtain a more accurate overall simulation time estimation.

Figure 6 shows a basic sensor model implementation in SystemC that communicates through TLM transactions (the *transport* method) to demonstrate the modeling principles. We will describe the example in more detail later in Sect. 6.1.

3   RISC-V Based VP Architecture

The VP is implemented in SystemC and designed as extensible and configurable platform around a RISC-V RV32IM CPU core with a generic bus system employing TLM 2.0 communication and support for the GCC toolchain—including coverage tracking with GCOV and debugging with GDB, of the software applications executed on our VP. Overall, the VP consists of around 3000 lines of C++ code with all extensions. Figure 1 shows an overview of the VP architecture. In the following we present more details.

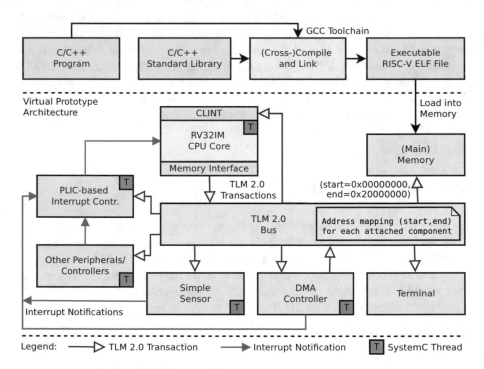

Fig. 1 Virtual prototype architecture overview

3.1 Core

The CPU core loads, decodes, and executes one instruction after another. We provide support for the RISC-V RV32IM instruction set in the CPU core and target the current version of the RISC-V machine level ISA as defined in the RISC-V privileged architecture specification [18]. This includes the machine level *control and status register* (CSRs) as well as instructions for interrupt handling (*wfi, mret*) and environment interaction (*ecall, ebreak*). We will provide more details on the implementation of interrupt handling and system calls (environment interaction) in the following sections.

3.2 Bus

The TLM bus is responsible for routing transactions from an initiator, i.e., (bus) master, to a target. Therefore, all target components are attached to the TLM bus at specific non-overlapping address ranges. The bus will match the transaction address with the address ranges and dispatch the transaction accordingly to the matching target. Please note, in this process the bus performs a *global-to-local* address translation in the transaction. For example, assume that a sensor component is mapped to the address range (start=0x50000000, end=0x50001000) and the transaction address is 0x50000010, then the bus will route the transaction to the sensor and change the transaction address to 0x00000010 before passing it on to the sensor. Thus the sensor works on local address ranges. The TLM bus supports multiple masters initiating transactions. Currently, the CPU core as well as the DMA controller are configured as bus masters. Please note that a single component can be both a master and a target, as for example the DMA controller receives transactions initiated by the CPU core to configure the source and destination address ranges and also initiates transactions by itself to perform the memory access operations without the CPU core.

3.3 Interrupts

Two sources of interrupts are available: (1) local and (2) external. Essentially, there are two sources of local interrupts: software and timer interrupts. Both are generated by the RISC-V specific *Core Local INTerruptor* (CLINT), which is configured through memory mapped I/O. External interrupts are processed with higher priority than local interrupts. External interrupts are all remaining interrupts triggered by the various components in the system. To handle external interrupts, we provide a PLIC-based *Interrupt Controller* (IC), based on the description from [18]. The IC will collect and prioritize all external interrupts and then route them to the CPU core one by one. We will describe the interrupt handling process in more details later.

3.4 VP Initialization

The main function in the VP is responsible to instantiate, initialize, and connect all components, i.e., set up the architecture. An ELF loader is provided to parse and load an executable RISC-V ELF file into the memory and set up the program counter in the CPU core accordingly. Finally, the SystemC simulation is started. The ELF file is produced by the GCC toolchain by (cross-)compiling the application program and optionally linking it with the C/C++ standard library (we also support a bare-metal execution environment without C/C++ library).

4   VP Interaction with SW and Environment

In this section we present more details on the HW/SW interaction, in particular on interrupt handling, and environment interaction via system calls in our VP.

4.1 Interrupt Handling and HW/SW Interaction

In the following we present an example application that periodically accesses a sensor to demonstrate the interaction between hardware (VP side) and software with a particular focus on interrupt handling. We first describe the software application running on the VP and then present a minimal assembler bootstrap code to initialize interrupt handling and describe how interrupts are processed in more details. Later in Sect. 6.1 we present the corresponding SystemC-based sensor implementation in our VP.

4.1.1   Software Side

Figure 2 shows an example application that reads data from a sensor and copies the data to a terminal component. The sensor and terminal are accessed through memory mapped I/O. Their addresses are defined at the top of the program. They need to match with the configuration in the VP. The sensor periodically triggers an interrupt, denoting that new data is available. The main function starts by registering an interrupt handler for the sensor interrupt (Line 27). Again, the interrupt number specified in SW has to match the configuration in the VP. Next, the sensor is configured in Line 29–30 using memory mapped I/O. The scaler denotes how

```
1  #include "stdint.h"
2  #include "irq.h"
3
4  static volatile char * const TERMINAL_ADDR = (char *
       const)0x20000000;
5  static volatile char * const SENSOR_INPUT_ADDR = (char *
       const)0x50000000;
6  static volatile uint32_t * const SENSOR_SCALER_REG_ADDR =
       (uint32_t * const)0x50000080;
7  static volatile uint32_t * const SENSOR_FILTER_REG_ADDR =
       (uint32_t * const)0x50000084;
8
9  _Bool has_sensor_data = 0;
10
11 void sensor_irq_handler() {
12    has_sensor_data = 1;
13 }
14
15 void dump_sensor_data() {
16    while (!has_sensor_data) {
17      asm volatile ("wfi");
18    }
19    has_sensor_data = 0;
20
21    for (int i=0; i<64; ++i) {
22      *TERMINAL_ADDR = *(SENSOR_INPUT_ADDR + i);
23    }
24 }
25
26 int main() {
27    register_interrupt_handler(2, sensor_irq_handler);
28
29    *SENSOR_SCALER_REG_ADDR = 5;
30    *SENSOR_FILTER_REG_ADDR = 2;
31
32    for (int i=0; i<3; ++i)
33      dump_sensor_data();
34
35    return 0;
36 }
```

Fig. 2 Example application running on the VP to demonstrate the hardware/software interaction

fast sensor data is generated and the filter setting what kind of post-processing is performed on the data. Finally, the copy process is iterated for three times (Line 32–33) before the program terminates. Each iteration starts by waiting for sensor data (Line 16–18). The global boolean flag *has_sensor_data* is used for synchronization. It is set in the interrupt handler (Line 12) and unset again immediately after the waiting loop (Line 19). Please note, the *wfi* instruction will power down the CPU core until the next interrupt occurs.

4.1.2 Bootstrap Code and Interrupt Handling

Figure 3 shows the essential parts of a bare-metal bootstrap code, which is written in assembler and linked with the application code, to handle interrupts.[2] The *_start* label is the entry point of the whole program. The registers *mtvec*, *mie*, and *mcause* are CSRs that essentially store the interrupt handler address, enabled interrupts, and interrupt source, respectively. The instructions *csrr* and *csrw* read and write a CSR into and from a normal CPU register, respectively. Before the main function is called (Line 10), the interrupt handler base address (level-0) is stored in *mtvec* (Line 6–7) and all interrupts are enabled (Line 8–9). After the main function returns, the VP simulation terminates, because a loop is detected which does not contain any further instructions (Line 13).

In general, an interrupt can occur at any time during execution of the application SW. All interrupts propagate to the interrupt controller (IC) first and are prioritized there. The CPU core only receives a notification that some interrupt is pending and needs to be processed. The CPU will first store the execution context, i.e., program counter and register values, and then read the base address from the *mtvec* CSR and set the program counter to that address, i.e., effectively directly jumping to the level-0 interrupt handler (first instruction at Line 16). The interrupt handler (level-

Fig. 3 Bare-metal bootstrap code demonstrating interrupt handling

```
1   .globl _start
2   .globl main
3   .globl level_1_interrupt_handler
4
5   _start:
6   la t0, level_0_interrupt_handler
7   csrw mtvec, t0
8   li t1, 0x888
9   csrw mie, t1
10  jal main
11
12  loop:
13  j loop
14
15  level_0_interrupt_handler:
16  csrr a0, mcause
17  jal level_1_interrupt_handler
18  mret
```

---

[2]Support for integration with the C/C++ library is also available, e.g., by executing the instructions at the beginning of the main function or integrating them directly into the *crt0.S* file, which is the entry point of the C library and similarly to the bare-metal code also calls the main function after performing some basic initialization tasks.

0) first in Line 16 reads the reason (i.e., local or external interrupt) for the interrupt into the *a0* CPU register, which according to the RISC-V calling convention [3] stores the first argument of a function call. Then in Line 17 an interrupt handler implemented in C is called (level-1, not shown in this example). Essentially, this level-1 handler deals with a local timer interrupt by resetting the timer and with an external interrupt by asking the IC for the actual interrupt number with the currently highest priority (through a memory mapped register access) and then calls the application provided interrupt handler function (Line 11–13 in Fig. 2, this step is ignored if none has been registered for the interrupt number). Finally, the *mret* instruction restores the previously stored execution context. Please note that storing and re-storing the register values can also be implemented in SW, by pushing and popping them to/from the stack before/after calling the level-1 handler, respectively.

4.2 Environment Interaction: Syscalls and C/C++ Library

System calls (syscalls) are executed by redirecting them to the host system running the VP simulation. This requires to pass arguments from the guest application into the host system and integrate the return values back into the guest application (i.e., memory of the VP). Implementing syscalls enables support for the C/C++ standard library. Furthermore, we can directly use GCOV to track the coverage of the applications simulated on our VP (the GCOV instrumentation requires syscall support to open and write to files).

For example consider the *printf* function provided by the C standard library. Most of its functionality is implemented as portable C code independent of the execution environment. Essentially, the *printf* function will apply all formatting rules and create a simple char buffer, which is then passed to the *write* system call. At this point interaction with the execution environment is required. Figure 4 shows the relevant part of a stub that is provided in the RISC-V port of the C library.[3] Essentially, the arguments of the system call are stored in the CPU registers *a0* to *a3* and the syscall number in *a7*. Then the *ecall* instruction is executed. The VP simulator will detect the *ecall* instruction and directly execute the syscall on the host system as shown in Fig. 5.[4] In case of the *write* syscall a pointer argument *buf* is passed. This is a pointer value from the guest system, i.e., an index in the VP byte memory array *mem*, and has to be translated to a host memory pointer in order to execute the *write* syscall on the host system. Therefore, the guest_to_host_pointer function (Line 5) adds the base address of the VP byte memory array, i.e., *mem* +

---

[3]Example based on the *newlib* port https://github.com/riscv/riscv-newlib.

[4]It is also possible to execute a trap handler, similar to the interrupt handler described in the previous section (e.g., essentially, jump to the level-0 interrupt handler with the *mcause* CSR being set to a syscall number), and then redirect the write to e.g. a terminal component.

```
1   #define SYS_write 64
2
3   ssize_t write(int fd, const void *buf, size_t count) {
4     return syscall(SYS_write, fd, (long)buf, count, 0);
5   }
6
7   long syscall(long n, long _a0, long _a1, long _a2, long _a3) {
8     // store arguments in CPU register and trigger ecall
9     register long a0 asm("a0") = _a0;
10    register long a1 asm("a1") = _a1;
11    register long a2 asm("a2") = _a2;
12    register long a3 asm("a3") = _a3;
13    register long a7 asm("a7") = n;
14
15    // special instruction causing a jump to the syscall handler
16    asm volatile ("ecall" : "+r"(a0) : "r"(a1), "r"(a2), "r"(a3),
          "r"(a7));
17
18    // store potential error code and return result
19    if (a0 < 0) {
20      errno = -a0;
21      return -1;
22    } else {
23      return a0;
24    }
25  }
```

Fig. 4 System call handling stub linked with the C library (guest side, executed on the VP host system)

```
1   #define SYS_write 64
2
3   // execute syscall on the host system
4   ssize_t sys_write(int fd, const void *buf, size_t count) {
5       const void *p = (const void *)guest_to_host_pointer(buf);
6       return write(fd, p, count);
7   }
8
9   long execute_syscall(long n, long _a0, long _a1, long _a2, long
        _a3) {
10      switch (n) {
11          case SYS_write:
12              return sys_write(_a0, (const void *)_a1, _a2);
13          //...
14      }
15  }
16
17  // function inside the CPU core
18  void execute_step() {
19    auto instr = mem_if->load_instr(program_counter);
20    auto op = decode(instr);
21
22    switch (op) {
23      case Opcode::ECALL: {
24        regs[a0] = execute_syscall(regs[a7], regs[a0], regs[a1],
            regs[a2], regs[a3]);
25      } break;
26      //...
27    }
28  }
```

Fig. 5 System call execution on the VP by redirecting to the host system

buf. The result of the syscall is stored in the *a0* register and passed back to the C library. We have implemented other syscalls in a similar way to the *write* syscall.

In general the guest and host system have a different architecture with different word sizes, e.g., in our case the guest system (which is simulated in the VP) is a 32 bit and the host system (which runs the VP) is a 64 bit system. Therefore, one has to be careful when data is passed between the guest and the host. Primitive types, e.g., int and bool, can be passed directly from the guest to the host, because our host system running the VP uses data types with equal or larger sizes, thus no information is lost when passing the arguments. When passing values back from the host a check can be performed, if necessary, to ensure that no relevant information is truncated, e.g., due to casting a 64 bit value into a 32 bit one. Pointer arguments need to be translated to host addresses, as described above, before accessing them on the host system. A write access is thus directly propagated back to the guest application. Structs can be accessed and copied recursively, considering the rules for accessing primitive and pointer types.

5   VP Performance Optimizations

In this section we discuss two performance optimizations for our VP that result in significant simulation speed-ups. The first optimization is a direct memory interface to fetch instructions and perform load/store operations from/to the (main) memory more efficiently. The second is a temporal decoupling technique with local time quantums to reduce the number of costly context switches, especially, in the CPU core simulation. We describe both techniques in the following.

5.1   *Direct Memory Interface*

The CPU core translates every load and store operation into a transaction which is routed through the bus to the target. Most of the time the main memory is the target of the access. Always accessing the memory through a bus transaction can be very costly. Even more so, because fetching the next instruction requires to load it from the memory too. Thus, at least one bus transaction is executed for every instruction. To optimize the access of the main memory and in particular instruction fetching, we provide two proxy classes with a direct memory interface. The direct memory interface stores the address offset where the memory is mapped in the overall address space as well as the size and pointer to the start of the memory. We have a proxy class for fetching instructions and one to access the memory in general, i.e., to perform load/store byte/half/word instructions. With the proxy classes enabled, the CPU core will first query the proxy class. It will match in case the main memory is accessed (for the instruction proxy class we only allow to fetch instructions from main memory) and otherwise convert the access into a transaction and normally route it through the bus.

5.2 Local Time Quantums

A SystemC-based simulation is orchestrated by the SystemC simulation kernel that switches execution between the various threads. While this is not a performance problem for most components, since they become runnable on very specific events, context switching can become a major bottleneck in simulating the CPU core. The reason is that a direct implementation will perform a context switch after executing every instruction, because simulation time has passed and the SystemC kernel needs to check for other runnable threads to perform synchronization. However, most of the time no other thread becomes runnable and the CPU thread is resumed again. Even if some other thread would become runnable, it is still fine to keep running the CPU thread for some time (ahead of the global simulation time of the system). For example, even if the sensor thread would be runnable and trigger an interrupt once executed, delaying the sensor thread execution for a small amount of time and keeping the CPU thread running should not have influence on the functional behavior of the system. In general the software does have no knowledge of the exact timing behavior and thus is written in such a way, e.g., by employing locks and flags, to always wait for certain conditions.

6 VP Extension and Configuration

Our VP is designed as a configurable and in particular extensible platform. It is very easy to add additional components (i.e., peripherals/controllers including bus masters) and attach them to the bus system at a new address range, or change the address mapping of the existing components. This allows for an easy (re-)configuration of the VP. By following the TLM 2.0 communication standard, transactions can be annotated with optional timing informations to obtain a more accurate timing model of the executed software. Support for additional RISC-V ISA extensions (beyond I and M) can be added inside the CPU core by extending the decode and execute functions accordingly. In general the compact implementation size (around 3000 lines of C++ code with all extensions) makes the VP very manageable and thus suitable as foundation for different application areas. In the following, we demonstrate the extensibility of our VP by two concrete examples: addition of a sensor peripheral and extension by GDB debug functionality from the application SW perspective.

6.1 Extending the VP with a Sensor Peripheral

This section presents the SystemC-based implementation of the VP sensor peripheral, which is used by the SW example presented in Sect. 4.1. It shows the
principles on modeling peripherals and extending our VP as well as demonstrates
the TLM communication and basic SystemC-based modeling and synchronization.
The sensor is instantiated in the main function of the VP alongside the other
components and attached to the TLM bus.

The sensor implementation is shown in Figs. 6 and 7. The sensor model has a data
frame of 64 bytes that is periodically updated (overwritten with new data, Line 34–
43) and two 32 bit configuration registers *scaler* and *filter*. The update happens in the
run thread (the run function is registered as SystemC thread inside the constructor
in Line 23). Based on the scaler register value this thread is periodically unblocked
(Line 30) by calling the notify function on the internal SystemC synchronization
event. Thus, *scaler* defines the speed at which new sensor data is generated. The
filter register allows to select some kind of post-processing on the data. After every
update an interrupt is triggered, which will propagate through the interrupt controller
to the CPU core up to the interrupt handler in the application SW. Therefore, the
sensor has a reference to the interrupt controller (IC, Line 4) and an interrupt number
provided during initialization (Line 20 and Line 21).

Access to the data frame and configuration registers is provided through TLM
transactions. These transactions are routed by the bus to the transport function
(Line 1). The routing happens as follows: (1) The sensor has a TLM target socket
field, which is bound in the main function (i.e., VP simulation entry point) to an
initiator socket of the TLM bus. (2) The transport function is bound as destination
for the target socket in the constructor (Line 22).

Based on the address and operation mode, as stored in the generic payload
(Line 2–3), the action is selected. It will either read (part of) the data frame
(Line 13) or read/write one of the configuration registers (Line 28–34). In case of a
register access a pre-read/write validation and post-read/write action can defined as
necessary. In this example, the sensor will ignore invalid scaler values (Line 21–25)
and reset the data generation thread on a scaler update (Line 37–40). Please note
that the transaction object (generic payload) is passed by reference and provides
a pointer to the data, thus a write access is propagated back to the initiator of
the transaction. Optionally, an additional delay can be added to the sc_time delay
parameter (also passed by reference) for a more accurate timing model.

6.2 Debugging Support Extension

We have implemented the GDB RSP (Remote Serial Protocol) interface to provide
direct debugging support of applications running on our VP with the GDB debugger
(in particular the freely available RISC-V port of the GDB, which knows about the

```
1   struct SimpleSensor : public sc_core::sc_module {
2     tlm_utils::simple_target_socket< SimpleSensor> tsock;
3
4     interrupt_controller *ic = 0;
5     uint32_t irq_number = 0;
6     sc_core::sc_event run_event;
7
8     // memory mapped data frame
9     std::array<uint8_t, 64> data_frame;
10
11    // memory mapped configuration registers
12    uint32_t scaler = 25;
13    uint32_t filter = 0;
14    std::unordered_map<uint64_t, uint32_t *> addr_to_reg;
15
16    enum { SCALER_REG_ADDR = 0x80,
17           FILTER_REG_ADDR = 0x84 };
18
19    SC_HAS_PROCESS(SimpleSensor);
20    SimpleSensor(sc_core::sc_module_name, uint32_t irq_number)
21        : irq_number(irq_number) {
22      tsock.register_b_transport(this, &SimpleSensor::transport);
23      SC_THREAD(run);
24      addr_to_reg = { {SCALER_REG_ADDR, &scaler},
25                      {FILTER_REG_ADDR, &filter} };
26    }
27
28    void run() {
29      while (true) {
30        run_event.notify(sc_core::sc_time(scaler,
             sc_core::SC_MS));
31        sc_core::wait(run_event); // 40 times per second by default
32
33        // fill with random data
34        for (auto &n : data_frame) {
35          if (filter == 1) {
36            n = rand() % 10 + 48;
37          } else if (filter == 2) {
38            n = rand() % 26 + 65;
39          } else {
40            // fallback for all other filter values: random
                  printable
41            n = rand() % 92 + 32;
42          }
43        }
44
45        ic->trigger_interrupt(irq_number);
46      }
47    }
48
49    //see Fig. 7 for transport function implementation
50  };
```

Fig. 6 SystemC-based configurable sensor model that is periodically filled with random data—demonstrates the basic principles on modeling peripherals

```
1   void transport(tlm::tlm_generic_payload &trans, sc_core::sc_time
         &delay) {
2     auto addr = trans.get_address();
3     auto cmd  = trans.get_command();
4     auto len  = trans.get_data_length();
5     auto ptr  = trans.get_data_ptr();
6
7     if (addr >= 0 && addr <= 63) {
8       // access data frame
9       assert (cmd == tlm::TLM_READ_COMMAND);
10      assert ((addr + len) <= data_frame.size());
11
12      // return last generated random data at requested address
13      memcpy(ptr, &data_frame[addr], len);
14    } else {
15      assert (len == 4);  // NOTE: only allow to read/write whole
           register
16
17      auto it = addr_to_reg.find(addr);
18      assert (it != addr_to_reg.end()); // access to non-mapped
           address
19
20      // trigger pre read/write actions
21      if ((cmd == tlm::TLM_WRITE_COMMAND) && (addr ==
           SCALER_REG_ADDR)) {
22        uint32_t value = *((uint32_t *)ptr);
23        if (value < 1 || value > 100)
24          return; // ignore invalid values
25      }
26
27      // actual read/write
28      if (cmd == tlm::TLM_READ_COMMAND) {
29        *((uint32_t *)ptr) = *it->second;
30      } else if (cmd == tlm::TLM_WRITE_COMMAND) {
31        *it->second = *((uint32_t *)ptr);
32      } else {
33        assert (false && "unsupported tlm command for sensor
             access");
34      }
35
36      // trigger post read/write actions
37      if ((cmd == tlm::TLM_WRITE_COMMAND) && (addr ==
           SCALER_REG_ADDR)) {
38        run_event.cancel();
39        run_event.notify(sc_core::sc_time(scaler, sc_core::SC_MS));
40      }
41    }
42  }
```

Fig. 7 Transport function for the SystemC-based sensor peripheral class (see Fig. 6)

available register set and the CSRs). Our VP acts as server and the GDB as client. They communicate through a TCP connection and send text based messages. A message is either a packet or a notification (a simple single char "+") that a packet has been successfully processed. Each packet starts with a "$" char and ends with a "#" char followed by a two digit hex checksum (the sum over the content chars modulo 256). For example the packet $m111c4,4#f7 has the content m111c4,4 and checksum f7. The m command denotes a memory read, in this case read 0x4 bytes starting from address 0x111c4. Our server might then for example return +$05000000#85, i.e., acknowledge the packet and return the value 5 (two chars per byte). To handle the packet processing and TCP communication we added a gdb-stub component to our VP. The whole debugging extension is only about 500 additional lines of C++ code, most of them to implement the gdb-stub. On the VP side, only the CPU core has been modified to lift the SystemC thread into the gdb-stub, to allow the CPU to interrupt and exit the execution loop in case of a breakpoint and thus effectively transfer execution control to the gdb-stub.

Debugging works as follows: Start our VP in *debug-mode* (command line argument), this will transfer control to the gdb-stub implementing the RSP interface, waiting for a connection from the GDB debugger. In another terminal start the GDB debugger. Load the same executable ELF file into the GDB (command "file main-elf") as in our VP. Connect to the TCP server of the VP (command "target remote :5005", i.e., to connect to localhost using port 5005). Now the GDB debugger can be used as usual to set breakpoints, continue, and step through the execution. It is also possible to directly use a visual debugging interface, e.g., *ddd* or *gdb-dashboard* or even the *Eclipse* IDE. Figure 8 shows a screenshot of debugging the sensor application in Eclipse.

Please note, the ELF file contains information about the addresses and sizes of the various variables in memory. Thus, a print(x) command with an int variable x is already translated into a memory read command (e.g., m11080,4). Therefore, on the server side, i.e., our VP, an extensive parsing of ELF files is not necessary to add comprehensive debugging support. In total we have only implemented 24 different commands of which 9 can simply return an empty packet and a few more some pre-defined answer. Relevant packets are for example: read a register (p), read all registers (g), read memory range (m), set/remove breakpoint (Z0/z0), step (s), and continue (c).

7   Experiments

In this section we present a performance comparison of our RISC-V based VP implementation with the RISC-V based *PULPino* platform (RTL implementation). For this comparison, the *PULPino* platform is simulated in a commercial RTL simulator. The *PULPino* platform is configured to use the *RISCY* core, which similar to our core also supports the RV32IM instruction set. We also demonstrate the

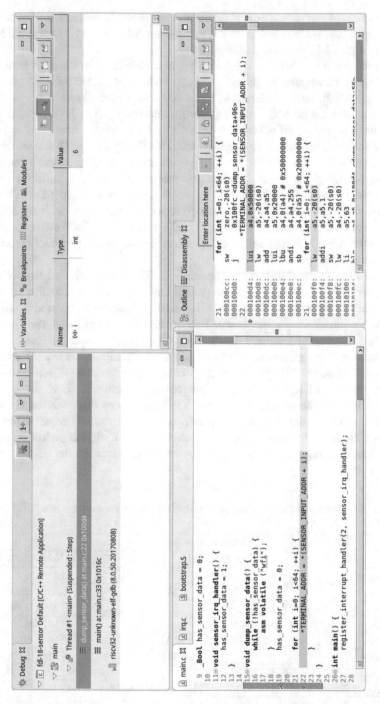

Fig. 8 Debugging the sensor application with Eclipse (screenshot showing relevant part of the debug view inside the Eclipse IDE)

effectiveness of our presented VP simulation performance optimization techniques. All experiments are performed on a Linux system with an AMD Opteron 2220 SE processor with 2.8 GHz and 32 GB RAM.

For the evaluation we use the following benchmarks from the RV8 benchmark set: (1) *primes* computes prime numbers up to a limit of 33,333,333; (2) *qsort* sorts an array with 50 million elements; (3) *sha512* applies the sha512 cryptographic hash function on a 60 MB data set (1 million iterations). The RV8 benchmark set contains some additional benchmarks, which we have omitted from the comparison due to problems on executing them on the PULPino platform in the commercial RTL simulator. In addition to the RV8 benchmarks, we have added a bubblesort (sorting 50,000 elements) and a recursive mergesort (sorting 1 million elements) benchmark to the comparison. Due to timeouts (set to 4 h, denoted T.O.) in the RTL simulation we also added down-scaled versions of the benchmarks (/s appended) to the comparison: primes/s has a limit of 33,333; qsort/s sorts 5000 elements; sha512 operates on a 0.6 MB data set (1000 iterations); mergesort/s and bubblesort/s sort 1000 elements, respectively.

Table 1 shows the results of the experiments. The table is divided in two halves: the upper half shows the down-scaled benchmark versions, while the lower half shows the longer running versions. All execution times are reported in seconds. The first column shows the benchmark name. The second and third columns show the number of executed instructions (measured on our VP) and LOC of the benchmark, respectively. The fourth column (PULPino) shows the execution time for running the benchmark on the PULPino platform (RTL implementation) in the commercial RTL simulator. The remaining columns show the execution time for running the benchmark on our VP with various optimization techniques enabled: no optimization (column: basic), with an instruction proxy using direct memory interface (dmi) for instruction fetching (column: +i_dmi), additionally with a data access proxy using dmi for loading/storing data from/to the (main) memory

Table 1 Experiment results: all execution times are reported in seconds, timeout (T.O.) set to 4 h (14,400 s)

Benchmark	#instr-exec.	LOC	PULPino RTL Sim.	Our RISC-V VP basic	+i_dmi	+d_dmi	+q10	+q100	+q1000
Bubblesort/s	2022052	20	787.49	0.97	0.71	0.58	0.43	0.41	0.40
Mergesort/s	297226	41	56.70	0.39	0.35	0.35	0.32	0.29	0.29
Primes/s	4341572	24	823.11	1.73	1.01	0.96	0.59	0.49	0.48
Qsort/s	290765	146	64.50	0.40	0.35	0.34	0.31	0.30	0.27
Sha512/s	8120416	154	1307.23	3.23	1.87	1.57	0.90	0.71	0.68
Bubblesort	200197558	20	T.O.	69.21	44.16	30.71	16.31	11.97	11.35
Mergesort	535918604	41	T.O.	197.32	107.89	86.48	41.17	27.77	26.15
Primes	7114988801	24	T.O.	2400.34	1214.71	1089.36	542.46	387.09	374.33
Qsort	3061611834	146	T.O.	1204.98	698.50	510.70	262.93	162.73	154.93
Sha512	8071548963	154	T.O.	2773.60	1556.02	1302.75	616.10	432.52	406.34

(column: +d_dmi), additionally with a local time quantum of 10 (column: +q10), 100 (column: +q100) and 1000 (column: +q1000) instruction cycles, respectively.

It can be observed that every optimization technique significantly improves the simulation performance on our VP. We observed a factor of improvement between 6.1x and 7.8x on this benchmark set with all optimization techniques enabled on the longer running benchmarks. Increasing the time quantum further beyond 1000 instruction cycles has only a minor effect on the simulation performance, because the impact of the SystemC thread context switch becomes marginal on the overall execution time. It can be observed that our VP is multiple orders of magnitude faster than the RTL simulation, especially, with optimizations enabled. Our VP executes up to 20 million instructions per second on the longer running benchmarks (between 17.6 and 20.5 million, depending on the benchmark) on our evaluation system (AMD 2.8 GHz).

8   Conclusion

In this paper, we have proposed and implemented the first RISC-V based VP to further expand the RISC-V ecosystem. The VP has been implemented in SystemC and designed as extensible and configurable platform around a RISC-V RV32IM core with a generic bus system employing TLM 2.0 communication. Our VP is very compact, with around 3000 lines of C++ code including all extensions, making it very manageable and thus suitable as foundation for various application areas, including early SW development and analysis of interactions at the HW/SW interface of RISC-V based systems. Finally, our RISC-V VP is fully open source to stimulate further research and development of ESL methodologies.

For future work we consider two different directions: (1) further extend our VP with additional components and RISC-V ISA extensions, and (2) verify our VP using verification techniques for SystemC, e.g., [8].

Acknowledgements This work was supported in part by the German Federal Ministry of Education and Research (BMBF) within the project CONFIRM under contract no. 16ES0565 and by the University of Bremen's Central Research Development Fund and by the University of Bremen's graduate school SyDe, funded by the German Excellence Initiative.

References

1. Bailey, B., Martin, G., & Piziali, A. (2007). *ESL design and verification: A prescription for electronic system level methodology*. Morgan Kaufmann/Elsevier
2. Binkert, N., Beckmann, B., Black, G., Reinhardt, S. K., Saidi, A., Basu, A., et al. (2011). The gem5 simulator. *SIGARCH Comput. Archit. News, 39*(2), 1–7. DOI: https://doi.org/10.1145/2024716.2024718. URL http://doi.acm.org/10.1145/2024716.2024718.

3. Calling convention. https://riscv.org/wp-content/uploads/2015/01/riscv-calling.pdf. Accessed: 2018-05-13.
4. Große, D. & Drechsler, R. (2010). *Quality-driven SystemC design*. Springer
5. Grüttner, K., Görgen, R., Schreiner, S., Herrera, F., Peñil, P., Medina, J., et al. (2017). CONTREX: Design of embedded mixed-criticality CONTRol systems under consideration of extra-functional properties. *Microprocessors and Microsystems, 51*, 39–55.
6. Herdt, V., Le, H. M., Große, D. & Drechsler, R. (2016). *On the application of formal fault localization to automated RTL-to-TLM fault correspondence analysis for fast and accurate VP-based error effect simulation - a case study*. In *FDL* (pp. 1–8).
7. Herdt, V., Le, H. M., Große, D., & Drechsler, R. (2017). *Towards early validation of firmware-based power management using virtual prototypes: A constrained random approach*. In *FDL* (pp. 1–8).
8. Herdt, V., Le, H. M., Große, D., & Drechsler, R. (2018). Verifying SystemC using intermediate verification language and stateful symbolic simulation. TCAD. DOI: https://doi.org/10.1109/TCAD.2018.2846638.
9. IEEE Std. 1666. (2011). IEEE Standard SystemC Language Reference Manual.
10. Leupers, R., Schirrmeister, F., Martin, G., Kogel, T., Plyaskin, R., Herkersdorf, A., & Vaupel, M. (2012). Virtual platforms: Breaking new grounds. In *DATE* (pp. 685–690).
11. Onnebrink, G., Leupers, R., Ascheid, G., & Schürmans, S. (2016). Black box ESL power estimation for loosely-timed TLM models. In *SAMOS* (pp. 366–371). DOI: https://doi.org/10.1109/SAMOS.2016.7818374.
12. RISCV-QEMU. https://github.com/riscv/riscv-qemu. Accessed: 2018-05-13.
13. RV8. https://rv8.io. Accessed: 2018-05-13.
14. Schuster, T., Meyer, R., Buchty, R., Fossati, L., & Berekovic, M. (2014). Socrocket - A virtual platform for the European Space Agency's SoC development. In *ReCoSoC* (pp. 1–7).
15. Spike. https://github.com/riscv/riscv-isa-sim. Accessed: 2018-05-13.
16. Streubühr, M., Rosales, R., Hasholzner, R., Haubelt, C., & Teich, J. (2011). ESL power and performance estimation for heterogeneous MPSoCs using SystemC. In *FDL* (pp. 1–8)
17. Waterman, A., & Asanović, K. (2017). The RISC-V Instruction Set Manual; Volume I: User-Level ISA. SiFive Inc. and CS Division, EECS Department, University of California, Berkeley.
18. Waterman, A., & Asanović, K. (2017). The RISC-V Instruction Set Manual; Volume II: Privileged Architecture. SiFive Inc. and CS Division, EECS Department, University of California, Berkeley.

AADD-Based Symbolic Simulation of SystemC AMS

Carna Zivkovic and Christoph Grimm

1 Introduction

Electronic design automation is based on modeling languages that allow designers to describe models of hardware or software. Examples are VHDL, Verilog, and SystemC (based on C++), just to name a few. The main purpose of such modeling languages is simulation.

However, other use cases such as synthesis and formal verification require formal models like automata or binary decision diagrams (BDDs) that represent the complete behavior of a model. The straightforward way to get such models is to write yet another compiler. Besides the huge effort for writing a compiler for languages such as C++ or VHDL, this is likely to introduce limitations to subsets, and maybe inconsistencies to the simulation results. The most significant disadvantage of two separate compilers or tools is that close integration of simulation with other use-cases can provide them with useful information, e.g. for concolic (mixed concrete/symbolic) verification or synthesis based on simulation-in-the-loop.

In this paper we show a way to use the existing proof-of-concept implementation of SystemC (AMS) for concolic simulation, based on polymorphism and operator overloading. In particular we show how a formal model can be created with a simple code-instrumentation without using an additional compiler. This paper extends the paper [18] with a more comprehensive description of the AADD formal model used for concolic simulation.

The concrete use-case is symbolic simulation of SystemC (AMS). In the following, we first give an overview of related approaches to generate formal models from modeling or programming languages with a focus on SystemC. In Sect. 2, a

C. Zivkovic (✉) · C. Grimm
TU Kaiserslautern, Kaiserslautern, Germany
e-mail: zivkovic@cs.uni-kl.de; grimm@cs.uni-kl.de

© Springer Nature Switzerland AG 2020   135
T. J. Kazmierski et al. (eds.), *Languages, Design Methods, and Tools for Electronic System Design*, Lecture Notes in Electrical Engineering 611,
https://doi.org/10.1007/978-3-030-31585-6_8

detailed description of AADD incl. its definition and operations is given. Section 3 explains code instrumentation to permit symbolic execution of C++ and symbolic simulation of SystemC (AMS). Section 4 introduce the methods for generating AADD using the method based on block condition tracking. In Sect. 5, we discuss the integration of the approach in different models of computation of SystemC (AMS). In Sect. 6, we give two SystemC AMS examples. In Sect. 7, we close the paper with a summary and conclusion.

1.1 State of the Art and Related Work

The straightforward way to translate a modeling language into a formal model is to write a compiler. For example, SystemC is translated by a compiler into intermediate representations like IVL [9], UPPAAL's timed automata [7], or simple sequential C [6]. A comprehensive overview including the particular challenges and limitations is given in [12].

The need of a separate compiler can be overcome by frameworks such as Java or the Clang C++ compiler that offer well-specified, open intermediate representations: the Java virtual machine, or the LLVM. This approach is taken by SystemC-clang [8] and PINAVM [11]. However, still a number of limitations remain. This includes e.g. the use of C++ pointers and loops (see [11, 12] for details).

In particular for testing and verification it is very useful to combine concrete and symbolic techniques (concolic testing [16]) in a single framework. Object-oriented features of modern languages permit e.g. [1, 4, 13] the symbolic execution within a regular compiler or simulator. In PyExZ3 [1], Python programs are translated to the input language of the automated theorem prover Z3. For this purpose, Python's integer objects, operators and functions are replaced by a symbolic type that produces the Z3 input language. In [4, 13], SystemC operators and functions are overloaded to permit symbolic simulation.

The approach taken in PyExZ3 [1] supports control flow by overloading functions that implement conditional statements. Compared with [1], we introduce a method that also handles the non-functional control-flow statements of procedural languages: block condition tracking. This allows us to generate AADD in a simple yet efficient way.

2 Symbolic Simulation of SystemC with AADD

2.1 Overview of Tool Flow

This work describes the language-related part of a toolkit for verification of mixed-signal systems. The toolkit targets the concolic (mixed concrete/symbolic) verification based on SystemC. To increase verification coverage, simulation of critical parts is done in a symbolic way. To improve interoperability with existing SystemC models and verification infrastructure, and to allow designers to deal with scalability issues, we permit the concrete simulation of the other parts. The tool flow is shown in Fig. 1.

The tool flow consists of the following steps:

1. Modeling: Models are specified in regular SystemC (AMS and/or TLM). By code-instrumentation a designer specifies which parts of a model are simulated symbolically, and which in a concrete way.
2. Compile & link: The model is compiled with a C++ compiler. After that, it must be linked with libSystemC (the SystemC simulator) and libAADD that provides symbolic extensions.
3. Execute: The resulting executable file is executed which starts a concrete simulation run. Where instrumented, the simulator will do a symbolic simulation run that

   (a) Follows all feasible paths in a comprehensive way,
   (b) Generates an AADD [5, 13] as formal model.

4. Check assertions: within (symbolic/concrete) simulation, assertions as described in [14] are checked, and reports are generated.

We use AADD (Affine Arithmetic Decision Diagrams) as a formal model because AADD are an extremely efficient representation of reachability for the given use case. However, the same approach can also be used to generate other internal representations such as control/data-flow graphs for the purpose of high-level synthesis.

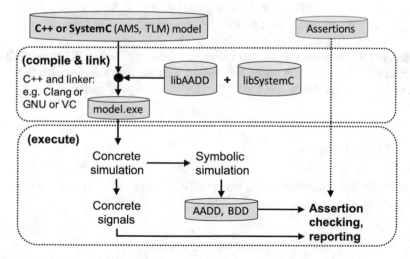

Fig. 1 Tool flow in which libAADD is applied

The libAADD is available on GitHub under https://github.com/TUK-CPS/ AADD. Some of the examples from this article are also provided there.

2.2 Internal Representation: AADDs

Affine Arithmetic Decision Diagrams (AADDs) combine two representations: Affine forms that represent the reachability in continuous variables, and binary decision diagrams that represent discrete (path) conditions.

The intuition behind AADD is illustrated by the short C++ program below. We will use it as a didactic example throughout the paper.

C++ Program with AADD

```
1:   #include "aadd.h" // symbolic types
2:   doubleS a(0,2); // a is from [0,2]
3:
4:   int main() {
5:     ifS(a > 1)
6:       a = a + 1;
7:     elseS
8:       a = a - 1;
9      endS
10:  }                    // a is now an AADD.
```

To distinguish symbolic and concrete semantics, we denote types and keywords that take symbolic values by a capital S at the end. The variable a, in the example, is a symbol for an unknown value from the range $[0, 2]$. Depending on the path condition $(a > 1)$, the then- or the else-part is executed. As a is from $[0, 2]$, both the then- and the else-part are feasible. After termination, a is from either $[-1, 0]$ or $(2, 3]$, depending on the path condition $(a > 1)$.

An AADD represents this information in a (reduced, ordered) binary decision diagram [2] whose internal nodes are labeled with the path conditions, and whose leaf nodes are affine forms that model the ranges and dependencies.

2.3 Affine Forms

An affine form $\tilde{x}$ represents an unknown value x from a range by a linear model of its dependency:

$$\tilde{x} ::= x_0 + \sum_{i=1}^{n} x_i \varepsilon_i.$$

The coefficient x_0 is the *center value*, the coefficients $x_1 \ldots x_n$ are the *partial deviations*. The *noise symbols* ε_i are symbolic variables that are restricted to $[-1, 1]$. Different affine forms (e.g. $\tilde{x} = x_0 + x_1\varepsilon_1$, $\tilde{y} = y_0 + y_1\varepsilon_1$) can share noise symbols. This represents the dependencies between them. Geometrically, the joint range of k affine forms is a zonotope in a k-dimensional space that is generated by the vectors of shared noise symbols, and centered around the vector of the center values.

Linear operations on affine forms and constants $c \in \mathbb{R}$ are defined as:

$$\tilde{z} \leftarrow c(\tilde{x} \pm \tilde{y}) ::= c(x_0 \pm y_0) + \sum_{i=1}^{n} c(x_i \pm y_i)\varepsilon_i.$$

Non-linear operations $\tilde{z} \leftarrow f(\tilde{x}, \tilde{y})$ are over-approximated by an affine form $f^a(\varepsilon_1, \ldots, \varepsilon_n)$ that reasonably well models f, and a new term $z_k\varepsilon_k$ that guarantees inclusion of the "real" value of z represented by $\tilde{z}$:

$$\tilde{z} \leftarrow f(\tilde{x}, \tilde{y}) \subseteq f^a(\varepsilon_1, \ldots, \varepsilon_n) + z_k\varepsilon_k.$$

For the non-linear operations, Chebychev approximations compute the coefficients while minimizing z_k (see [17] for details and other approximations).

The *fundamental invariant* of affine arithmetic [17] states that, at any instant between affine arithmetic operations, there is a single assignment of values from $[-1, 1]$ to each of the noise variables in use that makes the value of every affine form $\tilde{x}$ equal to the true value of the corresponding ideal quantity x.

2.4 Affine Arithmetic Decision Diagrams

Definition 1 (AADD) An AADD $\hat{x}$ represents the dependency of an unknown quantity $x \in \mathbb{R}$ from n noise variables $\varepsilon_1, \ldots, \varepsilon_n$ by a directed acyclic graph $(Q, T, E, \mathbb{X})$ with internal nodes Q, leaf nodes T, edges E, conditions $\mathbb{X}$, and it holds:

- Conditions $\chi_i \in \mathbb{X}$ are of the type $\chi_i ::= f^a(\varepsilon_1, \ldots, \varepsilon_n) \oslash 0$, where $\oslash$ is a relational operation.
- Internal nodes $v \in Q$

  - have two leaving edges $e_0, e_1 \in E$ that lead to child nodes $0(v), 1(v) \in T \cup Q$, respectively.
  - are labeled with $index(v)$; each $index(v) = i$ corresponds to a condition $\chi_i \in \mathbb{X}$.

- AADDs are ordered: for all edges (v_i, v_j) from $v_i \in Q$ to $v_j \in Q$: $index(v_i) < index(v_j)$
- Leaf nodes $v \in T$ are labeled with an affine form $aaf(v) = \tilde{x} = x_0 + \sum_{i=1}^{n} x_i\varepsilon_i$ with $n \in \mathbb{N}$.

To define the function that represents the true value of an AADD, we use the ITE (if-then-else) function. The ITE function takes three parameters: a condition $\chi \in \mathbb{X}$, and two AADDs. If χ is *true*, the result is the first AADD, otherwise the second.

Definition 2 (Value of an AADD) The value of the quantity $x \in \mathbb{R}$ represented by an AADD $\hat{x}$ with root $v \in Q \cup T$ is:

$$value(v) ::= \begin{cases} v \in T : \tilde{x} = x_0 + \sum_{i=1}^{n} x_i \varepsilon_i \\ v \in Q \text{ with } i = index(v) : ITE(\chi_i, value(1(v)), value(0(v))). \end{cases}$$

Remarks

- Both affine forms and BDDs can be considered as special cases of AADDs. Affine forms are AADDs that consist of just a leaf node. BDDs are AADDs whose leaves take only values from $\{0, 1\}$, or any other subset of the reals chosen to encode the Boolean values *true* and *false*. Furthermore, internal nodes of BDDs are labeled with binary inputs instead of the more general conditions $\mathbb{X}$ of AADD.
- Similar to affine forms, AADDs represent dependencies by sharing noise variables. In AADDs, noise variables also represent dependencies between the affine forms at the leaves and the conditions $\mathbb{X}$, because all of them (can) use the same noise variables $\varepsilon_1, \ldots, \varepsilon_n$. Even more, two AADDs can also share conditions by using internal nodes with the same index.
- Geometrically, an AADD with z leaves represent a set of z polytopes that are defined by:
  - z zonotopes that are generated by the affine forms at the leaves (linear dependencies).
  - The linear constraints introduced by the conditions (dependencies of conditions).

In the following, we describe operations on AADDs that ensure that, before and after each operation, there is a single assignment of values from $[-1, 1]$ to each of the noise symbolic variables that makes the value of each AADD $\hat{x}$ equal to the true value of the quantity $x \in \mathbb{R}$. This property is analog to the fundamental invariant of affine arithmetic.

2.5 Arithmetic and Relational Operations on AADDs

We describe arithmetic operations on AADDs in a recursive way, starting from the root nodes of the operands. Internal nodes of the result get the union of all conditions of both operands' internal nodes. Leaf nodes take the result of an affine operation on the respective leaf nodes of the operands.

Definition 3 (Arithmetic Operations on AADDs) Let $\hat{x}$, $\hat{y}$ be two AADDs with root nodes v_x, $v_y \in T \cup Q$, respectively. Arithmetic operations $\hat{x} \odot \hat{y}$ with $\odot$: $AADD \times AADD \rightarrow AADD$ are recursively defined:

1. For $v_x, v_y \in T$, the result is an AADD that is a leaf node v with $aaf(v) = aaf(v_x) \odot aaf(v_y)$.
2. For $v_x \in T$, $v_y \in Q$, the result is an AADD with root v and $index(v) = index(v_y)$ and $0(v) = v_x \odot 0(v_y)$ and $1(v) = v_x \odot 1(v_y)$.
3. For $v_x, v_y \in Q$ the result is an AADD with root v and, depending on the indices:
   If $index(v_x) = index(v_y)$: $index(v) = index(v_x)$, $0(v) = 0(v_x) \odot 0(v_y)$, $1(v) = 1(v_x) \odot 1(v_y)$
   If $index(v_x) < index(v_y)$: $index(v) = index(v_x)$, $0(v) = 0(v_x) \odot v_y$, $1(v) = 1(v_x) \odot v_y$
   If $index(v_x) > index(v_y)$: $index(v) = index(v_y)$, $0(v) = v_x \odot 0(v_y)$, $1(v) = v_x \odot 1(v_y)$.

In the following, we describe the relational operations on AADDs. We define relational operations with 0 as the right operand. Other relations can easily be transformed into this representation. We first define relations of affine forms of the form $\tilde{x} \oslash 0$ with $\oslash \in \{<, >, \leq, \geq, =, \neq\}$. If $0 \notin [lb(\tilde{x}), ub(\tilde{x})]$ we get a certain result from $\{true, false\}$. Table 1 shows these cases, and the equalities. Otherwise, the result of the comparison is uncertain and depends on the noise variables $\varepsilon_1, \ldots, \varepsilon_n$. In this case, we add the condition χ_k to $\mathbb{X}$, where k is an index that is not in use by any AADD.

For relational operations on AADDs, we have to consider that AADDs may have multiple leaves. The result of a relational operation is a binary decision diagram (BDD).

Table 1 Relational operations on affine forms of type $\tilde{x} \oslash 0$

	Cases for Boolean results		Otherwise: condition symbol
$\tilde{x} < 0$	$true$	: $ub(\tilde{x}) < 0$	$\chi_k \leftarrow \tilde{x} < 0$
	$false$	: $lb(\tilde{x}) \geq 0$	
$\tilde{x} \leq 0$	$true$	: $ub(\tilde{x}) \leq 0$	$\chi_k \leftarrow \tilde{x} \leq 0$
	$false$	: $lb(\tilde{x}) > 0$	
$\tilde{x} > 0$	$true$	: $lb(\tilde{x}) > 0$	$\chi_k \leftarrow \tilde{x} > 0$
	$false$	: $ub(\tilde{x}) \leq 0$	
$\tilde{x} \geq 0$	$true$	: $lb(\tilde{x}) \geq 0$	$\chi_k \leftarrow \tilde{x} \geq 0$
	$false$	: $ub(\tilde{x}) < 0$	
$\tilde{x} = 0$	$true$	: $ub(\tilde{x}) = lb(\tilde{x}) = 0$	$\chi_k \leftarrow \tilde{x} = 0$
	$false$	: $(ub(\tilde{x}) < 0) \vee (lb(\tilde{x}) > 0)$	
$\tilde{x} \neq 0$	$true$	: $(ub(\tilde{x}) < 0) \vee (lb(\tilde{x}) > 0)$	$\chi_k \leftarrow \tilde{x} \neq 0$
	$false$	: $ub(\tilde{x}) = lb(\tilde{x}) = 0$	

Definition 4 (Relational Operations on AADD) Let $\hat{x}$ be an AADD with root $v \in T \cup Q$, and m the highest index in use. Relational operations $\hat{x} \oslash 0$ with $\oslash \in \{<, >, \leq, \geq, =, \neq\}$ and $\oslash : AADD \times \{0\} \to BDD$ are defined as follows:

- For $v \in T$ with $aaf(v) = \tilde{x}$, the result is given by Table 1:
  Boolean results: a BDD that is a leaf node with value $true$ or $false$
  Otherwise: a BDD with root v_B and $index(v_B) = m + 1$, $0(v_B) = false$, and $1(v_B) = true$; $\chi_{m+1} = aaf(v) \oslash 0$ is added to the conditions' set $\mathbb{X}$.
- For $v \in Q$ the result is a BDD with root v_B and $index(v_B) = index(v)$, $0(v_B) = 0(v)$, $1(v_B) = 1(v)$ and $\chi_{index(v_B)} = \chi_{index(v)}$.

Remarks

- BDD result of relational operations on affine forms is of the height 1: the root node v_B is labeled with a condition $\chi_{index(v_B)} \in X$, and two leaves with the values $true$ and $false$.
- Relational operations on an AADD $\hat{a}$ result in a BDD $\hat{b}$ and for each $v \in Q$ of $\hat{a}$ with $index(v)$, there is at least one v_b of $\hat{b}$ with the same index.
- The result of arithmetic operations on AADDs $\hat{z} \leftarrow \hat{x} \odot \hat{y}$ has for each $v_x, v_y \in Q$ of $\hat{x}, \hat{y}$ with $index(v_x)$ and $index(v_y)$, resp. at least one $v_r \in Q$ of $\hat{z}$ with the same index.

2.6 Path Conditions as Linear Constraints on the Noise Variables

Leaf nodes $v \in T$ of AADDs are labeled with affine forms $aaf(v)$ that model the linear dependency from the noise variables. However, the range of the affine forms is further restricted by some conditions $\chi \in \mathbb{X}$ that also depend from the noise variables. The computation of upper and lower bounds that consider the conditions is an LP problem: Let $\mathbb{X}_P \subseteq \mathbb{X}$ be a set of the conditions $\chi_i \in \mathbb{X}$ on the path from the root node of an AADD to a leaf node with an affine form $\tilde{x}$. Then, the computation of accurate bounds of $\tilde{x}$ is an LP problem. For the upper bound $ub(\tilde{x})$:

$$\max(x_0 + x_1\varepsilon_1 + \ldots + x_n\varepsilon_n) \text{ subject to}$$

$$\mathbb{X}_P \text{ and} - 1 \leq \varepsilon_i \leq 1 \; \forall i \in \{1 \ldots n\}.$$

For the lower bound $lb(\tilde{x})$:

$$\min(x_0 + x_1\varepsilon_1 + \ldots + x_n\varepsilon_n) \text{ subject to}$$

$$\mathbb{X}_P \text{ and} - 1 \leq \varepsilon_i \leq 1 \; \forall i \in \{1 \ldots n\}.$$

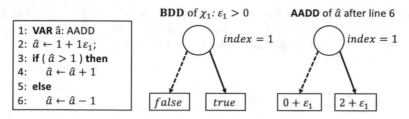

Fig. 2 Program (left) and the AADD $\hat{a}$ after the program run (right)

Example Figure 2 left shows the pseudo code of the short C++ program of the example given in Sect. 2.2, and right the AADD $\hat{a}$ after its execution. The condition $\hat{a} > 1 \Leftrightarrow \varepsilon_1 > 0$ depends on the unknown noise variable ε_1. We use the index 1 for this condition, hence $\chi_1 = (\varepsilon_1 > 0)$. χ_1 is represented by a BDD as shown in Fig. 2.

After line 6, for *true* value of the condition χ_1 the AADD $\hat{a}$ has a leave with an affine form $2 + \varepsilon_1 \in [1, 3]$ for $\varepsilon_1 \in [-1, 1]$. However, the condition $\chi_1 = (\varepsilon_1 > 0)$ restricts the range of ε_1 to $(0, 1]$ and hence, we get a tighter interval of $\hat{a}$ ($(2, 3] \subseteq [1, 3]$) for $\chi_1 = true$ after line 6. Same holds for $\chi_1 = false$.

2.7 Can AADD Be Reduced?

The advantage of AADD formal representation is that the size of the decision diagram can, in many cases, be reduced. The rules for reduction of AADD are similar to those applied on BDD [2], with some subtle differences:

Reduction of AADD Reduced AADDs can be obtained from AADDs by recursively applying the following reduction rules until none of them can be applied anymore:

1. Merge duplicate internal nodes with same indexes and same children; $v_1, v_2 \in Q$ with $index(v_1) = index(v_2)$, $1(v_1) = 1(v_2)$, $0(v_1) = 0(v_2)$ can be replaced with $v \in Q : index(v) = index(v_1)$, $1(v) = 1(v_1)$, $0(v) = 0(v_1)$.
2. Merge duplicate leaf nodes; $v_1, v_2 \in T$ with same values $aaf(v_1) = aaf(v_2)$ can be replaced with one leaf node $v \in T : aaf(v) = aaf(v_1)$.
3. Remove node with two identical children; $v \in Q$ with $1(v) = 0(v)$ can be replaced with $1(v)$.

The transformations (1) and (3) are done in the same way as for BDDs. In BDDs the transformation (2) is applied comparing Boolean values of leaf nodes. Leaf nodes with equal Boolean values are merged in one leaf node. For AADDs affine forms in leaf nodes are compared and two nodes are merged if affine forms are equal.

Definition 5 (Equality of Two Affine Forms) Let $\tilde{x} = x_0 + \sum_{i=1}^{n} x_i \varepsilon_i$ and $\tilde{y} = y_0 + \sum_{i=1}^{n} y_i \varepsilon_i$ be two affine forms. Two affine forms are equal if their center values are equal $x_0 = y_0$ and it holds that $\forall i \in \{1, \ldots, n\} x_i = y_i$.

3 Symbolic Execution of C++

3.1 Operator Overloading and Polymorphism

C++ and many other object-oriented languages support polymorphism and the overloading of operators and functions. For symbolic execution of C++ or symbolic simulation of SystemC, we provide a class AADD. This class can be used mostly like the C++ types double, float, and int. Furthermore, we provide the class BDD that can be used like the C++ type bool.

The class BDD implements an ROBDD [2]. In a symbolic execution run, an object of the class BDD holds all possible Boolean values of a variable of the type bool at its leaves, and the path condition in the internal nodes. The class BDD provides, among others, the following methods:

- Constructors from values of types bool and BDD,
- Logic operators and functions as defined on bool, but with BDD as parameters and result.
- Overloaded assignment operators, see Sect. 4.

The operators implement the known semantics of reduced ordered BDD.

The class AADD implements an AADD. An object of the class AADD holds a sound abstraction of values of a variable of the types int, float, or double in a symbolic execution. It provides among others the following methods:

- Constructors from values of int, float, double, AADD, and from interval bounds.
- Constructors from values of type bool and BDD; they convert bool and BDD to AADD replacing Boolean values *false* and *true* with 0 and 1.
- Arithmetic operators and functions as defined for int, float, and double.
- Relational operations with result of type BDD.
- Overloaded assignment operators, see Sect. 4.

In addition, the class implements methods for printing by overloaded stream operators, and an interface to the GLPK solver that is used to compute accurate interval bounds within symbolic execution, as explained in Sect. 2.6. The overloaded assignment operator is important for the handling of conditional statements. We describe it in Sect. 4. To hide the implementation, and to permit a more readable code we define:

Definition of Symbolic Data Types

```
typedef class AADD doubleS;
typedef class AADD floatS;
typedef class AADD intsS;
typedef class BDD  boolS;
```

The definitions above allow us to compute symbolically within a simulation run of an untouched SystemC simulator. We only have to include the header file that provides operators, functions, and methods with the appropriate signatures. Then, we can instantiate symbolic classes for symbolic execution, and use the expression- and compound-statements [10] of C++. This is shown by the didactic example given in Sect. 2.2.

3.2 Concrete and Symbolic Execution

We allow the designer to select parts that are executed with concrete values, and parts that are executed with symbolic values. This is useful as symbolic execution by principle suffers from the path explosion problem. To deal with this issue, one must carefully select critical parts for which symbolic execution is done. Switching semantics to concrete execution is also useful for debugging and in case of limited support for symbolic semantics (e.g. pointer arithmetic, etc.).

Selection between concrete and symbolic execution is done globally by the type of variables, e.g. double for concrete and doubleS for symbolic execution. Furthermore, we provide the macros CONCRETE and SYMBOLIC. These macros give the designer a more fine-grain control:

- CONCRETE(s, c) assigns a symbolic variable s a concrete value $c \in s$.
- SYMBOLIC(s, a) assigns a symbolic variable s a safe abstraction $a \supseteq s$.

Refinements resp. abstractions can either be chosen by specifying a value resp. a range, or by giving it constrained random values. The above macros also do the following checks:

- The macro CONCRETE checks if a given concrete value is a valid refinement of an AADD.
- The macro SYMBOLIC checks if a given symbolic AADD is a safe abstraction of a concrete value.

This mechanism allows us to combine symbolic and concrete execution.

For illustration, we use the didactic example. As we have not yet introduced the symbolic execution of selection statements, we may use the SYMBOLIC and CONCRETE macros in lines 4b and 6b. With these macros, we can model the didactic example as follows:

Use of CONCRETE and SYMBOLIC Macros

```
1: #include "aadd.h" // symbolic extensions
2: doubleS a(0,2); // a is from [0,2];
3:
4: int main() {
4b:   CONCRETE( a, 1.5 ); // concrete test
5:     if(a > 1)  a = a + 1;       // is executed a is 2.5
6:     else a = a - 1;     // this not.
6b:   SYMBOLIC( a, doubleS(1,3) );
7: } //continues with [1,3] if a in [1,3]
```

4   Block Condition Tracking

4.1   Code Instrumentation

The challenge with control flow statements is that for symbolic simulation we have to execute all feasible paths, while the concrete C++ control flow statements execute exactly one path. For example, in a selection statement (if-then-else), C++ will execute either the if or the then part, but not both.

In functional languages, where selection is a function, we can overload it by a function that

- Computes the condition (1st parameter)
- Computes the then-part (2nd parameter)
- Computes the else-part (3rd parameter)
- Returns an AADD or a BDD that has a new level with the 1st parameter as condition of a new internal node, and the 2nd and 3rd parameters as its child nodes.

However, procedural languages like C++ don't allow us to overload selection or iteration statements. Even worse, selection statements and iterations are no functions, which requires additional precautions to consider possible side effects. In the following, we overcome this limitation with a method we call "block condition tracking".

4.1.1   Selection Statements with Symbolic Semantics

Selection statements in C++ have the following syntax [10]:

Selection Statement

```
1: if (condition)
2:    statement  // then-part, arbitrary stmt.
```

```
3: [else        // optional:
4:    statement] // else-part
```

For block condition tracking, we instrument this statement. The instrumentation can be done manually or by a simple pre-processor provided with the AADD toolkit. An instrumented selection statement has the following syntax:

Instrumented Selection Statement

```
1: ifS (condition)
2:    statement
3: [elseS
4:    statement] endS
```

where

- condition is a C++ expression of type BDD.
- ifS(condition) is a macro that is executed at the beginning of a selection statement,
- elseS before the start of the else-part, and
- endS after the end of a selection statement.

We explain the function of the macros in the next section.

4.1.2 Iterations with Symbolic Semantics

Iteration statements are instrumented in a similar way as selection statements. In C++, a (while) iteration statement has the following syntax [10]:

Iteration Statement

```
1: while (condition)
2:    statement // arbitrary stmt.
```

By manual instrumentation, or by a preprocessor we translate it to an instrumented iteration statement that uses the macros whileS and endS. Again, these macros introduce code-fragments for symbolic execution of the iteration:

Instrumented Selection Statement

```
1: whileS (condition)
2:    statement; endS
```

4.2   Block Condition Tracking

In order to generate AADDs and BDDs by the instrumented selection and iteration statements, we track block conditions that are a subset to the path conditions.

A *path condition* is the conjunction of all conditions in a program run's conditional statements and iterations, from the program start to the current point of execution. In symbolic execution, we represent the path conditions by the decision tree in BDD or AADD. The tree is reduced immediately, if a path condition can be evaluated to either *true* or *false*.

A *block condition* is the conjunction of all conditions in nested selection or iteration statements. After an assignment, a block condition becomes part of the path condition of a variable.

We compute block conditions by pushing all conditions from selection and iteration conditions on a stack (stack of BDDs). Statements in a block are feasible (reachable) if the conjunction of block conditions is not false.

The instrumentations ifS, elseS, and endS track the block conditions:

- ifS pushes its parameter cond on a stack.
- elseS negates the condition on top of this stack, and
- endS pops the condition from the stack.

The block condition is then always the conjunction of all conditions on the stack. Note, that the instrumentation does not implement a selection statement. It only tracks the block condition. In consequence, both the statements in the if-part and in the else-part will be executed sequentially.

As an example, consider the selection statement of the didactic example from Sect. 2.2, with instrumentation:

Selection Statement in the Didactic Example

```
5: ifS(a > 1)
6:    a = a + 1;
7: elseS
8:    a = a - 1;
9: endS
```

With the macro ifS in line 5, the condition $a > 1$ is put on the stack. The macro does nothing else; it does not start an if-statement. Line 6 is therefore executed, independent from the condition. The macro elseS in line 7 negates the condition on top of the stack to $!(a > 1)$. Then, line 8 is executed, independent from the condition, and endS pops the condition from the stack. After line 8, a is an AADD that has the condition $a > 1$, a true-leaf $a + 1$, and a false-leaf $a - 1$, as shown in Fig. 2.

4.3 Building AADD and BDD by Overloaded Assignments

To build AADDs and BDDs we use the ITE function. The function ITE(cond, t, e) adds new levels to the decision diagrams. It has the following parameters:

- *cond* of type BDD,
- *t*, an AADD or BDD; the result for *cond == true*,
- *e*, an AADD or BDD; the result for *cond == false*.

The ITE function is implemented as follows (pseudo-code):

Implementation of ITE Function for BDD

```
1: FUNCTION: ITE(cond, t, e: BDD) returns BDD
2:    if (cond == true) then return t;
3:    if (cond == false) then return e;
4:    return (cond & t) | (!cond & e);
```

Note, that the parameters *t* and *e*, are AADDs or BDDs. The conjunction and disjunction functions for BDDs [2] merge the conditions for all tree parameters. The resulting BDD has all levels of the parameters and a new level for the condition cond. The algorithm is similar for AADD. However, we use (arithmetic) multiplication with 0 for false and 1 for true instead of conjunction, and addition instead of disjunction. Replacements of *true* and *false* values with resp. 1 and 0, are done by AADD constructor from BDD value.

We are now in a situation where we globally know all block and path conditions. Block conditions are on a stack of all possible "branches": so far, conditions in selection and iteration statements; later we handle symbolic process activations in the same way. Path conditions are represented in the decision diagrams of BDDs and AADDs.

We use this information in overloaded assignment operators. For an assignment lval := rval, concrete execution semantics will create a clone of rval and assign it to lval. In symbolic execution semantics, we have to merge the block conditions to the path conditions. This is done by the following method (in pseudocode):

Assign Method for AADD and BDD

```
0: global: stack of conditions of type BDD.
1: METHOD assign (rval, lval: AADD, BDD):
2:    bc := AND(all conditions on stack);
3:    rval := ITE(bc, rval, lval);
4:    return rval;
```

For illustration, we come back to the didactic example from Sect. 2.2. Lines 5–9, after expansion by the preprocessor become:

Expansion of Lines 5–9 in the Didactic Example

```
5:{ blkConds().thenBlock(a > 1);
6:    a = a + 1; // calls a=ITE(a>1,a+1, a);
7:    blkConds().elseBlock(__LINE__,__FILE__);
8:    a = a - 1; // calls a=ITE(!a>1,a-1, a);
9: blkConds().endBlock(__LINE__,__FILE__);}
```

Instead of an ifS-keyword in line 5, the condition $(a > 1)$ is put on the stack for the block condition. As there is no if-statement anymore, no matter which result the condition has, all following statements line 6–8 are executed and an AADD is generated.

Iteration statements are handled in a mostly similar way. However, a real iteration is executed, and hence the block condition `cond` in the previous iteration is popped from the stack and the block condition `cond` in the new iteration is put on the stack of block conditions by the `whileBlock` method. Note, that for termination of the loop we explicitly compare the condition of type BDD with false. When leaving the iteration statement, the block condition in the last iteration is popped from the stack. The code executed by the macros `whileS-endS` is then:

Execution of Loop Statement for AADD and BDD

```
while ((cond)!=false) {
    blkConds().whileBlock(cond);
    statement;
} blkConds().endBlock(); // pops the last condition.
```

Limitations

We have implemented and tested the selection statement and the iteration statement as described above. The above approach is not limited to these control-flow statements. Implementations of other syntactic forms of iteration and selection like do.. while, case.. select, for (...) are straightforward.

A bit ugly is the following issue: The transformations $if \rightarrow ifS, else \rightarrow elseS$ and $while \rightarrow whileS$ can be done easily e.g. by the C-preprocessor. However, inserting of $endS$ at the end of selection statements requires a parser that recognizes selection statements. We implemented such a preprocessor based on ANTLR by adding approx. ten lines to a listener class of its CPP14 grammar.

A more fundamental issue is that AADD use safe abstractions that are meaningless for e.g. pointer arithmetic and representation of bit-vectors.

5 Symbolic Simulation of SystemC (AMS)

5.1 Symbolic Signals and Process Activation

For symbolic simulation of SystemC (AMS), we have to consider the impact of symbolic representations on the signal representations, and on the activation of processes. Therefore, we first formalize some aspects of signals and the underlying models of computation (MoC).

Signals Let a signal be a sequence of samples $\langle v(t_1), v(t_2), \ldots, v(t_n) \rangle$, where the $t_i, i \in \mathbb{N}$ are elements of the simulated time, and $v(t_i)$ is the value of the sample at time t_i.

Concrete and Symbolic Signals We call a signal concrete (or deterministic), if each $v(t_i)$ is a concrete value. We call a signal symbolic (or uncertain [5]) if at least one sample is symbolic, e.g. a BDD or an AADD, and represents more than one value.

A symbolic signal represents all possible signal trajectories. As consequence of the fundamental invariant of affine arithmetic, it holds that there is a single assignment of values from $[-1, 1]$ to each of the noise variables that makes the trajectory of the symbolic signal equal to the trajectory of a 'real', concrete signal. This is a fundamental difference to the flow-pipe representation of signals that only represents an enclosing hull.

Processes Let a process be specified by $P = (I, O, proc, a)$ with

- input signals I and output signals O,
- $proc$, a processing method,
- a, an activation condition.

$proc$ uses samples from the input signals I to compute samples of the output signals O, maybe using internal states. The execution of $proc$ is done at points in simulated time, when a is $true$.

Activation Conditions An activation condition is concrete (deterministic), if its values are either $true$ or $false$. An activation condition is symbolic (uncertain), if at least one of its values is a symbol, i.e. represented by a BDD or an AADD.

5.2 Symbolic Simulation with Concrete Activation Conditions

In many relevant cases, we have concrete activation conditions. This depends on the models of computation, and is the case for:

- *Timed data flow (TDF)*. The TDF model of computation of SystemC AMS is based on the static data flow MoC. TDF/Static data flow defines a static schedule before start of simulation. It depends only on the rates of ports and modules, but not on other values. If rates are concrete, TDF have static activation conditions.

- *Continuous-time models (CT)*. The selection of analog solution points depends on the simulator. The objective of CT models is to minimize the quantization error between the unreachable ideal of infinitely many analog solution points, and a discrete number thereof. Hence, from a functional point of view, CT has a static activation condition.
- *Discrete-event (DE)*. The activation condition in DE MoC is the sensitivity list of SC_METHOD or wait statements of SC_THREAD. The activation condition is concrete, if the respective signals or events are based on concrete values (e.g. a concrete clock).

For concrete activation conditions, the process activation mechanism is not influenced by symbolic simulation results. For SystemC AMS, symbolic simulation requires hence no further modifications. For the SystemC DE MoC, we have concrete activation conditions if the sensitivity list consists of concrete signals. As an example, consider a counter that counts if a symbolic input th is above a threshold:

Counter Example

```
1:   sc_in<bool>    clk; // concrete signal
2:   sc_in<doubleS> th;  // symbolic signal
3:   sc_out<intS>   cnt; // symbolic signal
4:
5:   void count() {
6:      if (th>2) cnt += 1;
7:   }
8:
9:   SC_METHOD(count) sensitive << clk;
```

The above example has a concrete activation by the concrete signal clk. Unfortunately, SystemC's event detection mechanism is part of all DE signals, even if they are not part of the sensitivity list. We handle this by overloading a direct template instantiation.

5.3   Symbolic Simulation with Symbolic Activation Conditions

In the DE MoC, we can have the general case of a symbolic activation condition. The work in [19], presented at DATE 2019, shows the approach to handle activation conditions depending on symbolic signal values. Symbolic values of time are not considered and they are subject of future research. In [19] it is shown that in principle symbolic activation conditions can be treated like those for selection statements. We activate the process, but all assignments are made only under the true value of the activation condition a using ITE function. The process activation condition is a block condition of type BDD. The proposed implementation is as:

- Push a on the stack of block conditions.
- Execute the process; use ITE function for all assignments.

For more details we refer readers to [19].

6 Examples

6.1 Simple Example: Water Level Monitor

As a simple toy example, we model a water level monitor in SystemC AMS. We use the TDF model of computation. We model the water level by a variable `wlevel` of type `doubleS`. The water tank has two sensors that signal whether the water level is lower than 5 (port `l5` of type `boolS`) or greater than 10 (port `g10` of type `boolS`). The processing method called in each time step is for the water tank model:

Processing Method of the Water Level Tank Model

```
void processing() {
  if(pump)    // dynamics of water level
    wlevel+=(1.+uncertainty1)*timestep;
  else
    wlevel+=(-2.+uncertainty2)*timestep;

  if(wlevel < 5) l5 = true; // sensors
  else l5 = false;

  if(wlevel > 10) g10 = true;
  else g10 = false;
}
```

The controller's processing method switches a pump on, depending on the sensor's values (in-ports of type `boolS`):

Processing Method of the Controller

```
void processing() {
  if( l5 )  pump = true;
  if( g10 ) pump = false;
}
```

The SystemC model with its instrumentations for symbolic simulation is compiled by the regular C++ compiler (on OS X, LLVM), and linked against the

SystemC, SystemC AMS and AADD libraries. After running the executable, we get the well-known outputs, and in addition some reporting from libAADD:

Output of Executable File

```
        AADD lib -- Symbolic execution is enabled.
          AADD library (c) TU Kaiserslautern,
                   C. Zivkovic, C. Grimm.


       SystemC 2.3.0-ASI --- Apr 27 2017 16:27:00
       Copyright (c) 1996--2012 by all Contributors,
                  ALL RIGHTS RESERVED


        SystemC AMS extensions 2.0 Version: 1.0
              Copyright (c) 2010--2013  by
                  Fraunhofer-Gesellschaft
           Institut Integrated Circuits / EAS
      Licensed under the Apache License, Version 2.0

Info: SystemC-AMS:
3 SystemC-AMS modules instantiated
1 SystemC-AMS views created
3 SystemC-AMS synchronization
        objects/solvers instantiated

Info: SystemC-AMS:
1 dataflow clusters instantiated
cluster 0:
    3 dataflow modules/solver, contains e.g. module: wtank
    3 elements in schedule list, 100\,ms cluster period,
     ratio to lowest:  1 e.g.module: wtank
     ratio to highest: 1 sample time e.g. module: wtank
  0 connections to SystemC de,
  0 connections from SystemC de

Symbolic simulation took: 3.17225\,s.
```

The result of symbolic simulation is a sequence of AADD/BBD that represent all possible trajectories of the signals. To get a result that can be plotted and reasonably well understood we plotted the minimum and maximum values in a file. This is shown by Fig. 3.

6.2   Analog/Mixed-Signal Example: Delta-Sigma Modulator

As a more complex example we model a 3rd-order delta-sigma modulator. Like in the first example we use the timed data flow model of computation in SystemC AMS. The block diagram of the modulator is shown in Fig. 4.

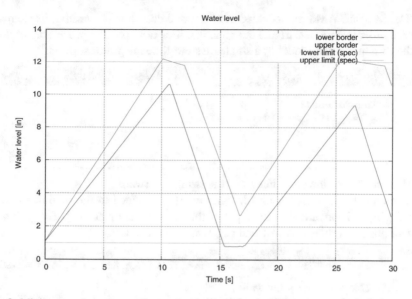

Fig. 3 Minimum and maximum of water level trajectories over time

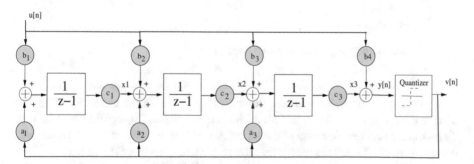

Fig. 4 Block diagram of 3rd order Delta-Sigma modulator [15]

The values of the coefficients are taken from [15]:

$$b_1 = 0.0444; \; b_2 = 0.2881; \; b_3 = 0.7997$$

$$a_1 = -0.0444; \; a_2 = -0.2881; \; a_3 = -0.7997$$

$$c_1 = c_2 = c_3 = 1; \; b_4 = 1.$$

The integrator outputs x_1, x_2, x_3 are computed by discrete-time integration of the integrator input signals.

The SystemC AMS model consists of a timed data flow cluster that implements the 3rd order integrator part and the quantizer. For example, the processing method of the SystemC AMS model of a single discrete-time integrator stage is:

Processing Method of a Single Discrete-Time Integrator

```
void processing() {
  x_1=x_1+0.0444*(u-v);
}
```

Implementation of the other integrators is straight-forward.

The output signal of the third integrator $x_3[n]$ is added to the modulator input signal $u[n]$ and forwarded to the input of the one-bit quantizer $y[n]$. The quantizer sets all positive values of y to 1 and negative values and 0 to -1.

SystemC AMS Module of the One-bit Quantizer

```
SCA_TDF_MODULE(quantizer) {
  sca_tdf::sca_in<doubleS> y;
  sca_tdf::sca_out<doubleS> v;
  void processing() {
    if(y>0) v=1;
    else v=-1;
  }
  quantizer(sc_module_name nm){}
}
```

By symbolic simulation and assertion checking we check the model for all inputs $u[n]$ in the range $[-0.5, 0.5]$, and all initial states $x_1[0]$, $x_2[0]$, $x_3[0]$ in $[-0.1, 0.1]$. To guarantee accurate function of the modulator, integrator saturation, and quantizer overload are checked. This means that the respective outputs of the integrators, and inputs of the quantizer must be in a range $[-2, 2]$. We verify this by symbolic simulations, where we assign the inputs a range (`doubleS(-0.5, 0.5)` and initial states resp. `doubleS(-0.1, 0.1)`.

After the symbolic simulation values of all signals are a sequence of AADD. Figure 5 plots minimum and maximum values of the integrator output x_3 for 20 time steps in one symbolic simulation run. The symbolic simulation for the given example took around 10.3 s. This includes the time for writing the results into a file which requires a solver call to compute upper/lower bounds for each point to be plotted.

For comparison, a single, concrete simulation run takes 2 ms. However, using random inputs, and given strongly nonlinear dynamic behavior of a quantizer, it is extremely unlikely to find the unknown corner cases.

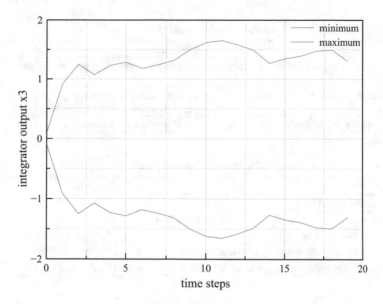

Fig. 5 Worst case signal values for x_3 at 20 time steps

7 Summary and Conclusion

7.1 Symbolic Simulation, Without Another Compiler?

The objective of the work was to avoid the implementation of a C++/SystemC compiler, and to permit a close interaction of concrete simulation with symbolic simulation.

By polymorphism and operator overloading, we can completely go without another C++ compiler for expression statements including function and method calls, and compound statements. For these kinds of statements the classes doubleS, floatS, intS, boolS provide sufficient functionality. The instrumentation of these types is necessary to distinguish concrete and symbolic semantics, e.g. by giving variables ranges of possible values.

Iteration and selection statements require macros that modify the control flow and track conditions. This has no semantic need. However, it is easy to introduce these macros via the keywords if, else, while by the preprocessor. Unfortunately, adding the macro endS at the end of an iteration or selection statement requires a simple compiler. We implemented it with 10 lines of code and ANTLR. Table 2 gives an overview.

Table 2 C++: what is needed without another compiler

C++ construct	Requires
Expression (statement), compound statement, declarations	Operator overloading, polymorphism
Selection statement, iteration statement	Instrumentation or own preprocessor adding ENDS, block condition tracking
Goto statement try statement (exceptions)	Likely as above, still not done

Table 3 SystemC: what is needed without another compiler?

SystemC MoC	Requires
AMS extensions: TDF	./.
DE with process activation on concrete time/value event	./.
AMS extensions: CT or LSF	Modified solver, e.g. [3]
DE with process activation on symbolic value event	(Proposed in [19])
DE with process activation on symbolic time event	(Future work)

Regarding simulation semantics of SystemC (AMS), concrete process activations are supported without any limitations. This is sufficient for the TDF MoC, and DE processes activated by a deterministic clock. Symbolic process activations are not supported in this work. The recent work [19] presented one possible solution for handling symbolic activation conditions with symbolic signal values. For this purpose, we combined the block condition tracking proposed in this paper with instrumentation of SystemC processes and overloading *update()* and *write()* methods of SystemC signals for handling symbolic signal values of types AADD, BDD. Symbolic values for points in time are still subject of future research. Table 3 gives an overview of required changes.

7.2   Only for Symbolic Simulation?

Although we worked with SystemC (AMS) and AADD in this paper, the approach is not limited to this particular use case at all. Just for example, control-data-flow graphs are a good starting point for high-level synthesis, or for timing analysis. The central idea of block condition tracking and overloaded assignments as a rather abstract method can easily be modified to generate control-data-flow graphs. For this purpose, the code that creates the internal representations has to be rewritten.

References

1. Ball, T., & Daniel, J. (2014). Deconstructing dynamic symbolic execution. In *Proceedings of the 2014 Marktoberdorf Summer School on Dependable Software Systems Engineering*. https:// www.microsoft.com/en-us/research/wp-content/uploads/2016/02/dse.pdf.
2. Bryant, R. E. (1986). Graph-based algorithms for boolean function manipulation. *IEEE Transactions on Communications, 35*(8), 677–691. http://dx.doi.org/10.1109/TC.1986.1676819.
3. Grabowski, D., Grimm, C., & Barke, E. (2006). Semi-symbolic modeling and simulation of circuits and systems. In *IEEE International Symposium on Circuits and Systems (ISCAS)* (pp. 983–986). Washington: IEEE. https://doi.org/10.1109/ISCAS.2006.1692752. http://ieeexplore. ieee.org/xpl/freeabs_all.jsp?arnumber=1692752.
4. Grimm, C., Heupke, W., & Waldschmidt, K. (2005). Analysis of mixed-signal systems with affine arithmetic. *IEEE Transactions on Computer-Aided Design of Integrated Circuits and Systems, 24*(1), 118–123. https://doi.org/10.1109/TCAD.2004.839469(410)24. https:// ieeexplore.ieee.org/document/1372667/.
5. Grimm, C., & Rathmair, M. (2017). Dealing with uncertainties in analog/mixed-signal systems: Invited. In *Proceedings of the 54th Annual Design Automation Conference, DAC 2017*, Austin, 18–22 June 2017 (pp. 35:1–35:6). https://doi.org/10.1145/3061639.3072949. http://doi.acm. org/10.1145/3061639.3072949.
6. Große, D., Le, H. M., & Drechsler, R. (2010). Proving transaction and system-level properties of untimed SystemC TLM designs. In *Proceedings of the Eighth ACM/IEEE International Conference on Formal Methods and Models for Codesign, MEMOCODE '10* (pp. 113–122). Washington: IEEE Computer Society. https://doi.org/10.1109/MEMCOD.2010.5558643.
7. Herber, P., Fellmuth, J., & Glesner, S. (2008). Model checking SystemC designs using timed automata. In *Proceedings of the 6th IEEE/ACM/IFIP International Conference on Hardware/Software Codesign and System Synthesis, CODES+ISSS '08* (pp. 131–136). New York: ACM. http://doi.acm.org/10.1145/1450135.1450166.
8. Kaushik, A., & Patel, H. D. (2013). Systemc-clang: An open-source framework for analyzing mixed-abstraction SystemC models. In *Proceedings of the 2013 Forum on Specification and Design Languages, FDL 2013*, Paris, 24–26 September 2013 (pp. 1–8). http://ieeexplore.ieee. org/document/6646649/.
9. Le, H. M., Große, D., Herdt, V., & Drechsler, R. (2013). Verifying SystemC using an intermediate verification language and symbolic simulation. In *2013 50th ACM/EDAC/IEEE Design Automation Conference (DAC)* (pp. 1–6). https://doi.org/10.1145/2463209.2488877. http://ieeexplore.ieee.org/document/6560709/.
10. Marchetti, A. *Hyperlinked c++ BNF grammar*. http://www.nongnu.org/hcb.
11. Marquet, K., & Moy, M. (2010). Pinavm: A systemc front-end based on an executable intermediate representation. In *Proceedings of the Tenth ACM International Conference on Embedded Software, EMSOFT '10* (pp. 79–88). New York: ACM. http://doi.acm.org/10.1145/ 1879021.1879032.
12. Marquet, K., Moy, M., & Karkar, B. (2009). A theoretical and experimental review of systemc front-ends. In *Forum on Specification and Design Languages 2009*. https://doi.org/10.1049/ic. 2010.0140. https://hal.archives-ouvertes.fr/hal-00495886.
13. Radojicic, C., Grimm, C., Jantsch, A., & Rathmair, M. (2017). Towards verification of Uncertain Cyber-Physical systems. *Electronic Proceedings in Theoretical Computer Science, 247*, 1–17. https://doi.org/10.4204/eptcs.247.1. https://doi.org/10.4204.
14. Radojicic, C., Grimm, C., Schupfer, F., & Rathmair, M. (2013). Verification of mixed-signal systems with affine arithmetic assertions. *VLSI Design, 2013*, 14. http://dx.doi.org/10.1155/ 2013/239064.
15. Sammane, G. A., Zaki, M. H., Tahar, S., & Bois, G. (2007). Constraint-Based verification of delta-sigma modulators using interval analysis. In *50th Midwest Symposium on Circuits and Systems* (pp. 726–729).

16. Sen, K., Marinov, D., & Agha, G. (2005). Cute: A concolic unit testing engine for C. In *Proceedings of the 10th European Software Engineering Conference Held Jointly with 13th ACM SIGSOFT International Symposium on Foundations of Software Engineering, ESEC/FSE-13*, pp. 263–272. New York: ACM. http://doi.acm.org/10.1145/1081706.1081750.
17. Stolfi, J., & de Figueiredo, L. H. (2003). An introduction to affine arithmetic. *TEMA Tendências em Matemática Aplicada e Computacional, 4*, 297–312. https://tema.sbmac.org.br/tema/article/view/352.
18. Zivkovic, C., & Grimm, C. (2018). Symbolic simulation of SystemC AMS without yet another compiler. In *Forum on Specification and Design Languages 2018*, pp. 5–16.
19. Zivkovic, C., & Grimm, C. (2019). Nubolic simulation of AMS systems with data flow and discrete event models. In *DATE 2019, Accepted for a Long Presentation; To be Presented in March 2019*

Blech, Imperative Synchronous Programming!

Friedrich Gretz and Franz-Josef Grosch

1 Introduction

Synchronous languages have in certain cases been successfully used for embedded software programming in industrial projects in the past [4] but they remain an exotic tool known only to experts. They are neither part of a general embedded programming curriculum nor are they used outside special, safety-critical embedded applications in industry. This is particularly regrettable because the key features of these languages, such as reactive, concurrent programming and the guarantee of causality, could mean a conceptual leap forward in the programming of most of today's embedded applications. Esterel [12] mostly targeted hardware design, not application-level software development. As of today its development has ceased and no up-to-date compiler is available. Quartz [14] can be regarded as a successor to Esterel but has the same hardware focus, limiting its applicability in software development. SCCharts [15] uses a graphical notation of hierarchical state charts and may be regarded as another successor to Esterel. It overcomes certain limitations of Esterel but its focus on graphical programming does not suit our needs as we briefly explain in Sect. 9. Scade [8] is an industrial-grade tool but it is tailored to safety-critical applications in particular industrial domains. Its dataflow orientation caters to its special target audience but may not be the best choice for many other embedded software programmers. Céu [13], a recently developed imperative language, aims at the programming of reactive embedded applications but it provides no causality guarantees. We therefore propose a new, imperative, synchronous, and

---

F. Gretz (✉) · F.-J. Grosch
Robert Bosch GmbH, Corporate Research, Renningen, Germany
e-mail: Friedrich.Gretz@de.bosch.com; Franz-Josef.Grosch@de.bosch.com

© Springer Nature Switzerland AG 2020
T. J. Kazmierski et al. (eds.), *Languages, Design Methods, and Tools for Electronic System Design*, Lecture Notes in Electrical Engineering 611,
https://doi.org/10.1007/978-3-030-31585-6_9

purely software-development oriented programming language called *Blech*.[1] From our point of view, the main benefit of synchronous programming is the automatic, causally correct composition of concurrent subprograms. At the same time, this is the single most challenging feature to accommodate for when designing the semantics and implementing compilers for such languages.

In the next section, we familiarise the reader with our language *Blech*. After that, we turn our focus on some key aspects regarding causality. We discuss prior work in Sect. 3. Subsequent sections describe our design choices and their consequences. Section 9 discusses a synchronous programming example in detail. We use the stopwatch example from [1] to reiterate several typical requirements of software engineering, stress the points we make about causality and separate compilation, and demonstrate how we imagine programming an embedded application.

2 Blech

The following listing shows a sample program written in *Blech*. It simply runs a PID controller[2] and toggles between two modes of operation depending on the user input. Despite its modest functionality, this program demonstrates different language features that may give an impression of what can be expressed in *Blech*. Below, we explain the relevant details.

```
enum Button
  Left default
  Right
end
                                                                          5
struct PID
  var KI: float32 = 0.0
  var KD: float32 = 0.0
  var KP: float32 = 1.0
  var dt: float32 = 1.0                                                   10
with
  function this:calc (pv: float32, sp: float32)
                     (priErr: float32,
                      intI: float32, cmd: int32)
    let error = sp - pv                                                  15
    let der = (error - priErr) / this.dt
    intI = intI + (error * this.dt)
    cmd = this.KP * error + this.KI * intI + this.KD * der
    priErr = error
  end                                                                    20
```

---

[1]Blech is German and colloquially translates to "bare metal" highlighting our focus on deeply embedded architectures.

[2]For a quick overview see [16].

```
   activity this:control  (pv: float32,  sp: float32)
                          (cmd: int32)
     var priErr: float32 = 0.0
     var intl: float32 = 0.0                                    25
     repeat
       this:calc(pv, sp)(priErr, intl, cmd)
       await true
     end
   end                                                          30
end

@[EntryPoint]
activity main () ()
   extern let sensor: float32                                   35
   extern var output: float32
   extern let pressed: event(Button)
   var pid: PID
   let sp: float32 = 42.17
                                                                40
   repeat
     cobegin
       await let b = pressed, b == .Left
     with weak
       pid.KP = 0.9; pid.KI = 0.0                               45
       run pid:control(sensor, sp)(output)
     end

     cobegin
       await let b = pressed, b == .Right                       50
     with weak
       pid.KP = 0.5; pid.KI = 0.5
       run pid:control(sensor, sp)(output)
     end
   end                                                          55
end
```

2.1 Reactions

A *Blech* program is triggered by the runtime environment at every tick of some external clock. The program responds to a tick by performing a *reaction step* which usually will read input data and compute output data from it. Following the synchrony assumption, the runtime environment must ensure that all input variables are sampled at the beginning of the reaction and remain unchanged while the reaction executes. The reaction starts in the entry-point activity *main* in line 34. Activities are one kind of subprograms in *Blech*. We also use functions and will explain the difference later. An activity will usually declare some local variables.

2.2 Declarations

Declarations are indicated by the `var` and `let` keywords. The former means that
the variable is indeed mutable whereas the latter declares immutable (or read-only)
data.[3] Additionally, `extern` tells the compiler that the variable is defined outside
this program and we can simply expect it to be in our scope after linking. This is
necessary whenever we receive data from the runtime environment or write data to
it as is the case with sensor readings and controller commands. Every declaration
must either provide a data type explicitly (as in lines 35–39) or it may be deduced if
the right-hand side of the declaration uniquely determines the data type. Also, every
declaration will automatically initialise a variable to the type's default value. This is
0 for numerical types, and in case of the *PID* structure the default value is given in
the type declaration (lines 7–10).

2.3 Events

A special, built-in generic data type is the `event` type. Events are a special form
of an optional data type wherein the event either is absent or it is present and
carries some payload. In order to become present during a reaction, an event must
be *emitted* either by the runtime environment if it is an external event, or by using a
special `emit` keyword for internal events. In any case, the runtime clears all events
at the end of a reaction so that an event does not persist from one reaction to another
unless it is re-emitted again for the next reaction. In line 37, we define an external
event which indicates that a user has pressed a button. The payload is of type *Button*,
defined by the enumeration in lines 1–4, and tells us which of the two buttons has
been pressed. The use of events will be explained in the next paragraph.

2.4 Statements

The *main* activity consists of an infinite `repeat..end` loop. In the body of that
loop, the control flow is forked into two branches using the `cobegin..with..end`
statement. Conceptually, i.e. from the programmer's point of view, the two branches
are executed concurrently. The compiler, however, will sequentialise these branches
into one sequence of instructions after an automated causality analysis, which
ensures that such a sequentialisation exists. Generally, the branches of a `cobegin`
statement will join when each of them has terminated. The `weak` keyword may be
used to indicate branches which may be aborted. Let us see how this works out in
lines 42–47. In line 43, the branch consists solely of an `await` statement. When the

---

[3]Here `let` should not be confused with the same keyword in functional languages where it binds
a free variable in a subsequent subexpression.

control flow reaches an `await` statement, the execution along that branch is stopped. A reaction is finished when every branch has hit an `await` statement. Upon the next tick, a new reaction starts and all branches resume execution from the `await` statements that they have ended in previously. Every `await` statement is equipped with some condition. Often this condition is simply `true` which means that the program just waits for a new trigger to continue (e.g. in line 28). The condition in line 43 is more complex. Remember that *pressed* is an event which may or may not be present, so the first part checks for presence and in such case copies the payload to the local immutable variable *b*. Then we check whether the payload's value is *Button.Left*.[4] If both the presence test and the comparison succeed, the execution continues beyond the `await` statement and in this case the branch terminates. If either fails, the control flow remains at the `await` statement and waits for the next reaction to start. Concurrently, in lines 45–46, some fields of the *pid* struct are reinitialised and then the *control* activity is started that operates on *pid* and some given arguments. The effect of making the `with` branch `weak` is that in the reaction where the user has pressed the left button, the concurrently running *control* activity is executed until it finishes its current reaction step and then completely terminated, the branches are joined and the control flow proceeds to the next statement in line 49. Essentially, this is a simple way to guard a repetitive, possibly infinite, behaviour with some abortion condition. Note that the same behaviour could have been expressed using a (non-immediate) `abort` statement. There exists a variety of synchronous preemptions, cf. [14, p. 34]. In *Blech* we support a selection of those and we will use them in the running example of Sect. 9. In general, `cobegin` allows any number of `with` blocks and any block can be made `weak`. Summing up, the *main* activity defines two modes of operation given by the setting of the PID controller in lines 45 and 52 and toggles between these states whenever the user presses the right or the left button.

2.5 Subprograms

Let us look in more detail at the different kinds of subprograms in *Blech*, namely functions and activities. Generally, subprograms will have a list of read-only parameters (inputs) and a list of read-write parameters (outputs). When calling a subprogram, all arguments are passed by reference. The difference to activities is that functions are *instantaneous*. This means that a function must run to completion during a reaction step. It may not use `await` statements or call activities. Functions in *Blech* will usually encapsulate computation instructions (as in lines 12–20) or complex expressions. Another use case for functions is to access or modify structured data in a consistent way. Activities, on the other hand, typically maintain some state information either in their local variables or in their control flow and

---

[4]By default enumerations are opened and hence their tags may be accessed without specifying the type name.

carry this state from one reaction step to another. This is useful when programming
the mode switching logic of an application component.

2.6 Type Extensions

Finally, let us turn our attention to the structure type defined in lines 6–31. In *Blech*,
any data type may be extended by additional static code artefacts such as constants
or methods. This example shows an extension with two methods: a function and an
activity. The subtle difference between top-level activities or functions and extension
methods is that methods always must be given a reference to an instance of the
data type that they extend. For example, in line 12 the identifier[5] *this* is used
to reference a given instance of the *PID* struct. A field of this instance can be
accessed as in `this.KP`. If we were to allow a method to not only read but also
modify the contents of an instance, then the method needs to be declared as a
`mutating function` or `mutating activity`. This informs the causality analysis
about the intended use of the passed instance reference. Code generation of methods
is rather simple because they can be rewritten as normal functions or activities that
simply receive an extra input or (when they are mutating) an extra output parameter.
The *PID* structure shows how we can organise data and code in an object-based
way. Currently we do not support features of object-oriented programming such as
inheritance, polymorphism through interface abstractions, or generics. The first two,
in their full generality, require the program to look up the correct implementation of
a method at run time. This is known as dynamic dispatch and imposes too large a
runtime penalty in an embedded real-time program. Generics would either require
a runtime representation with the aforementioned drawbacks or we would have to
generate monomorphic code for every instance of a generic data type (much in the
fashion of C++ templates). The latter solution then raises questions regarding the
intermediate representation of generic code to allow for separate compilation and
also how to globally analyse and minimise the number of monomorphic instances.
For the time being, we restrict the language to a set of built-in special container data
types which are frequently needed, such as arrays, optionals and events. Meanwhile,
we collect requirements for more general object-oriented features.

Generally, note that in *Blech* all data is allocated statically except for stack-
allocated local variables in functions. By design, there is no way to dynamically
allocate memory on the heap and then deallocate or garbage collect it. In the same
fashion, `cobegin` allows you to create an arbitrary but statically fixed number
of concurrent branches. It is not possible to create "worker threads" based on
some dynamic input. The reasons behind these decisions are that dynamic memory
management is error prone and in the case of garbage collection unpredictable in
terms of runtime. Moreover, in a safety-critical application one would expect to
have a guaranteed memory bound after compilation. In practice, even with this

---

[5]It is not a predefined keyword as in Java but an arbitrary identifier as in F#.

memory bound, it is already challenging to ensure that the application meets its timing constraints. Finally, we believe that these features are not required by the control-oriented, embedded real-time applications that we are targeting with *Blech*.

2.7 Causality

The PID example above uses concurrent composition of statements. However, in that particular example, the statements are unrelated in terms of data flow. One branch is observing whether a particular button has been pressed while the other branch deals with the calculation of controller commands from sensor readings. In general, it may be the case that the concurrent branches access shared data, which raises the question: in what order do they access that data and do they have a consistent view? In the next sections, we answer this question and discuss implications.

3 Key Questions and Related Work

Generally speaking, a program is causal if during a reaction step all concurrently running threads have a consistent view on shared data. A detailed definition of causally correct programs turns out to be non-trivial and gives rise to various notions of "constructive semantics"—a concise overview can be found in [9, Sect. 9]. Once a causality notion is fixed, related questions arise.

3.1 How Can Concurrent Calls to Subprograms Be Composed in a Causally Correct Way?

Consider the typical example taken from [5] with two programs which are represented as sets of equations:

$$P : \forall n \in \mathbb{N} \begin{cases} x_n = f(u_n) \\ y_n = g(v_n) \end{cases}$$

$$Q : \forall n \in \mathbb{N} \quad v_n = h(x_n)$$

The problem is whether P and Q can be composed concurrently. Alternatively, the program could be directly specified as

$$R : \forall n \in \mathbb{N} \begin{cases} x_n = f(u_n) \\ v_n = h(x_n) \\ y_n = g(v_n) \end{cases}$$

The difference is that in R all assignments happen concurrently and obviously a causally correct scheduling can be found. However, the concurrent composition of P and Q is not possible if P is sequentialised and compiled prior to composition with Q because then no causal order exists. Different approaches to separate compilation have been explored in literature:

Lublinerman et al. [11] propose a best-effort approach that "clusters" a subprogram into non-overlapping, concurrent parts. Thus, the intermediate compiled code of a subprogram consists of precompiled parts and scheduling constraints among these parts. This allows the composition of precompiled subprograms by interleaving those precompiled parts while respecting the scheduling constraints and causality constraints. In this way, the problem above is solved; however, the decomposition is not transparent to the programmer—changing the implementation may alter the compiler-generated decomposition and break existing software. Furthermore, the decomposition strategy requires the compiler to make trade-offs, e.g. between the amount of code duplication and reusability, which are beyond the programmer's control.

Benveniste et al. [5] propose the extraction of an interface out of the code of a given subprogram. This interface is represented as an automaton with different kinds of relations between its states. Based on such interface descriptions it can be decided whether two subprograms are concurrently composable and how their individual actions need to be scheduled to maintain causal order. Thus, the problem in the example above is solved in a similar manner as in [11] but with similar problems from a software-engineering point of view.

In Quartz, compilation is the transformation of code to an intermediate format: the so-called synchronous guarded actions. Scheduling and causality analysis are not part of the compilation [7]. Calling subprograms in Quartz amounts to copying the corresponding code wherein all names (formal parameters) are substituted by the supplied arguments. Hence, there is no difference between R and the concurrent composition of P and Q from the example above. The drawback, however, is that causality is a global property and is only decided in a final code synthesis stage. Modular software development is impossible because there is no interface to program against.

3.2   How Can Structured Data Types Be Used Concurrently?

Any program will usually make use of data structures commonly known as "structs" and "arrays". However, in synchronous programming these structures become problematic when passed into concurrent subprograms.

Looking at the C code generator for Quartz, we see that arrays are decomposed into individual variables that represent the cells of an array. The causality analysis is then straightforward but it also means that all array accesses must be evaluated at compile time. In other words: a for-loop running over an array is not implementable in Quartz.

In Scade [8], arrays permit only special, side-effect-free operations such *map* and *fold* known from functional languages. Other operations are outside the language and have to be implemented in the host language. This raises the issue of dealing with foreign function code in the causality analysis.

Recently, Aguado et al. [2] proposed the use of a variant of interface automata to define admissible operations on an encapsulated data object. Their theory could allow a programmer to wrap, for example, arrays in an object that provides getters and setters and a policy that ensures a causally correct usage. From a language designer's point of view, however, we may ask whether this approach permits too much. If everyone may define an arbitrarily complex usage policy for any object, is it feasible to use third-party code and understand the potential error messages when used incorrectly?

4 Causality

We choose acyclic schedulability as the causality notion that fits software needs. This results in simple programming rules:

- every variable is declared in the scope of a thread (which may read and write this variable arbitrarily once it is visible)
- upon a fork, such a variable may be shared between the subthreads of which at most one may read and write it; the others may only read the variable and only after the last writing operation of the writer has finished in the current reaction step
- upon joining, the original (parent) thread reclaims all its access rights

This is a special case of the sequentially constructive semantics for synchronous languages [10] and can also be seen as a synchronous implementation of a causal memory model [3]. Our notion of a statically determined writer and potential readers aligns well with recent developments in actor-based languages such as Rust.[6] or Pony[7] In today's embedded software, the lack of clearly determined reaction steps, and writers and readers within such steps, is one of the main pain points in embedded software development. To our knowledge, Scade also only implements acyclic schedulers [8] but, being a functional or dataflow-oriented language, it does not permit sequential read-write operations as we do.

Let us exemplify the programming rules for *Blech* using the PID controller example from Sect. 2. Assume we have two controller objects *pid1* and *pid2*. For the sake of argument, say that in each reaction the output of *pid1* is used as one of the inputs of *pid2*. This can be written in *Blech* as follows:

---

[6] www.rust-lang.org.

[7] www.ponylang.org.

```
// assuming variables in1, in2, sp, out1, out2 are in scope ...
cobegin
   run pid1:control(in1, sp)(out1)
with
   run pid2:control(in2, out1)(out2)
end
```

Our compiler will automatically deduce from the data flow of variable *out1* that in every reaction of this program, a control step of *pid1* has to be performed first, and only then a step of *pid2* is done. The lexicographic order of the cobegin branches is irrelevant for this sequentialisation, which means the following program behaves exactly the same:

```
cobegin
   run pid2:control(in2, out1)(out2)
with
   run pid1:control(in1, sp)(out1)
end
```

Using shared variables opens the door for two kinds of programming mistakes which both are automatically detected by the compiler. First, write-write conflicts:

```
cobegin
   run pid1:control(in1, sp)(out2)
with
   run pid2:control(in2, sp)(out2)
end
```

In the above example, both branches try to concurrently write to the same variable *out2* by mistake. This is forbidden. Compilation will stop and indicate the error to the programmer. Another possible mistake is introducing circular read-write dependencies as in the next example.

```
cobegin
   run pid1:control(in1, out2)(out1)
with
   run pid2:control(in2, out1)(out2)
end
```

It is impossible to execute this program because *pid2* needs the output of *pid1* which in turn requires the output of *pid2*. Again, our compiler stops with a corresponding error message. Sometimes control algorithms do have such feedback loops but they are never instantaneous. There must be a known value, usually a value from the previous reaction, that can be used to start the current reaction. In *Blech*, this is expressed using the prev operator.[8]

```
cobegin
   run pid1:control(in1, out2)(out1)
with
   run pid2:control(in2, prev out1)(out2)
end
```

---

[8]In Simulink, delays are used in the same fashion to resolve algebraic loops.

This program tells the compiler to take the previous value for *out1*. Thereby, the causality cycle is broken and sequential code can be successfully generated.

Automated causality analysis not only is a useful feature but also a guiding principle in *Blech*'s design. In the rest of the paper we focus on several of those design choices and explain them in more detail.

5 Separate Compilation

A common theme to all papers mentioned in the introduction is that the causality interface of a subprogram is the result of some static analysis. Our approach goes in the opposite direction: we ban global variables entirely and use two parameter lists for our functions and activities: a list of input (read-only) parameters and a list of output (read-write) parameters. In this way, it is the programmer who defines a simple causality interface for his subprograms. Thus, any program can be composed of precompiled subprograms which act as black-boxes and only declare a set of read-only and a set of read-write variables. On the basis of these interfaces alone, the compiler can check if a causally correct composition exists. This black-box approach enables us to treat causality interfaces as contracts in the same way as classical function interfaces: the programmer may alter the implementation as long as the new one abides by the same interface. The calling code never sees the change of the implementation. This is a crucial decision to allow for programming modular, maintainable and separately testable applications.

Reconsider the example from page 167: In *Blech*, P and Q would be activities. Variables u and v are input variables, x and y are output variables for P. Thereby, it is explicitly stated that in every reaction step both u and v are read to produce new values for x and y. Based on this information, our causality analysis will not allow the concurrent composition of P and Q. Indeed, if it really was the programmers intention to perform the two completely unrelated operations from P, he would be better off writing two activities: P_1 that computes x from u and P_2 that computes y from v. Then, P_1, P_2 and Q could be composed—or the programmer implements R directly. Based on our experience so far, we believe that this simple, explicit, no-compiler-magic approach to subprogram interfaces is the most viable in the long run, especially for large projects.

6 Structured Data

We distinguish two kinds of data types: primitive (or atomic) and structured. Examples of atomic types are all numeric types or enumerations. They all represent a piece of memory that contains one value. Structured data types, on the other hand, represent a chunk of memory that is subdivided into smaller partitions. For example, structs are divided into fields that are accessible by name while arrays are subdivided into cells that are accessible by indices. Causality analysis is clearly defined for atomic types but structured types require us to make a decision: how fine grained should the causality analysis be? Consider the following (toy) example where the control flow is forked into two concurrent branches which both try to write into shared memory:

```
struct Complex
  var real: float32
  var img: float32
end

function setReal (val: float32)(c: Complex)
  c.real = val
end

function setImg (val: float32)(c: Complex)
  c.img = val
end

  // ... in main activity ...
  var a: Complex
  cobegin
    setReal(-17.0)(a)
  with
    setImg(42.0)(a)
  end
```

We know from the interface of the functions *setReal* and *setImg* that the struct *c* may be modified. Hence, calling these functions concurrently with the same output argument *a* results in a potential write-write conflict (and is hence forbidden). This is despite the fact that the two functions would write into disjoint memory locations. As with subprograms, we take a black-box approach to causality analysis of data structures. So, by design, the causality interface does not specify which parts of a structured data type are read or written—it is always considered as a whole by the causality analysis. This ensures that the implementation does not "leak" into the interface. If we really want to concurrently write to disjoint parts of a structured data type, then the caller has the responsibility of determining these parts. This means the callee must be designed to receive only the part it is writing to. We can rewrite the above example in this manner:

```
// Complex as before
function setValue (val: float32)(loc: float32)
  loc = val
end

  // ... in main activity ...
  var a: Complex
  cobegin
    setValue(-17.0)(a.real)
  with
    setValue(42.0)(a.img)
  end
```

This is a valid program. Admittedly, the function *setValue* is not very interesting in this example but in practice it could encapsulate some validation logic which we do not want to repeat twice.

While our semantics may be regarded as a trivial special case of the framework in [2], we believe this simplicity is a great advantage. Structured data can always be shared among any number of readers but is owned by only one writer as a whole. If different parts need to be written in reactions to different events, then either the writer needs to know about all these events and react to all of them accordingly, or the data structure needs to be disassembled into disjoint parts that are given to different writers where each reacts to one event only.

7 References

Orthogonally to the notion of data being atomic or structured, we also discern between value- and reference-types. Note that this has nothing to do with the way the data is handled in the generated C code. For instance, value-typed arrays will nonetheless be passed around functions using pointers; the distinction between reference- and value-types is a semantic one—not an implementation-specific one. Primitive types, structs as well as fixed-sized arrays are value-types in *Blech* by default. Sometimes, however, references to data are needed. Typical use cases include

- Every formal parameter of a subprogram is a reference to a given argument.
- Aliasing of individual locations in a complex data structure, e.g.
  ```
  let ref rpm = wheels[3].rpm
  let ref rad = wheels[3].radius
  var ref speed = wheels[3].speed
  run rpmAsKmh(rpm, rad)(speed) // update the wheel's speed in
      every reaction given its rpm and radius
  ```

- Building data types that point to other data
  ```
  struct ValueStruct
    var a: int32
    var b: bool
  end

  ref struct RefStruct
    var x: float32
    var ref l: ValueStruct
  end
  ```

RefStruct contains a reference to a location that contains a value-typed *ValueStruct*. As such, *RefStruct* itself is a reference-type, indicated by the `ref` keyword in its declaration. These *reference-types* are necessary to structure code in an object-based way. References (i.e. aliases of other locations) as well as reference-type variables (i.e. instances of reference-typed data structures) must be assigned directly at their declaration and cannot be subsequently mutated. Note that their contents, however, may change throughout their lifetime—only the address is immutable.

```
var x: int8 = 5
var ref r = x
x = 17 // now also: r = 17
r = 42 // now also: x = 42
```

Immutability of references ensures that causality analysis remains decidable: we can statically determine the location that any given name points to. If we were to allow mutation, we would run into the undecidable problem of aliasing. The keywords `let ref` declare an immutable reference that may only read its contents while `var ref` declare an immutable reference that allows its contents to be changed. Our semantics require that `ref` is idempotent, i.e. a reference to a reference directly points to the original value, however, the access capabilities may change. Continuing the example above:

```
let ref s = r // s points to x, but read-only
var ref t = s // compiler error: cannot grant write access to
    read-only location
```

A particular feature is that we have no dereference and, accordingly, no address operator. Depending on the context that a reference is used in, the compiler automatically generates the correct code that either accesses the contents of a reference type or passes the address it is pointing to. Consequently, whenever a reference is expected (say as a function parameter) in *Blech* we can pass a value-typed variable or even a literal and the compiler will take care of finding the address or creating a temporary location.

Remember that in Sect. 2 we have introduced the `prev` operator to resolve causality cycles. It is important to note that this operator may only be used on value-typed data. The reason is that value-typed data (even structs or arrays) may efficiently be copied to store the previous value. References, on the other hand, will usually define a tree of objects which we would have to walk over to create a deep copy. In order to exclude this performance pitfall, we generally have the rule: "no prevs on refs!"

8 Sharing Data

So far, we have always assumed that an output parameter of a subprogram is distinct from any other formal parameter, i.e. the two names never point to the same memory location. This permits concurrent access to those parameters within the subprogram and also guarantees the programmer that writing an output parameter does not alter other parameters. On the calling side, this assumption restricts the caller since any two arguments must represent completely disjoint memory locations (unless they are both inputs). Consider the following activity which sums two numbers in every reaction and concurrently checks its second parameter for some threshold. When the threshold is exceeded, the weak branch is aborted and the whole cobegin block terminates.

```
activity add (a: int32, b: int32)(s: int32)
  cobegin
    await b > 10
  with weak
    repeat
      s = a + b
      await true
    end
  end
end
```

```
// ... in main activity ...
var sum: int32
var x: int32; var y: int32

run add(x, y)(sum)    // OK, all distinct
run add(x, x)(sum)    // OK, overlapping inputs
run add(x, sum)(sum)  // error
```

The last call uses *sum* in both the input and output lists. This is not possible in general, because our separate compilation will compile the activity without knowing how it will be used. Thus, the chosen sequentialisation of the cobegin-block may be fixed arbitrarily, in particular the check b > 10 may be done before an iteration of the loop. This is not a problem as long as *b* and *s* are pointing to disjoint memory locations. If they do not, as in the last call, the threshold check is done on *sum* before a new value for *sum* is computed, which is not causally correct.

Sometimes, however, the programs are less restrictive. For example, consider an activity that just adds two numbers.

```
activity add (a: int32, b: int32)(s shares a, b: int32)
  repeat
    s = a + b
    await true
  end
end
```

Here, it does not matter whether *a*, *b* and *s* represent the same location or not. The programmer explicitly declares this using the shares keyword. It is thereby guaranteed that these locations are not used concurrently inside the activity. Consequently, the call

```
run add(x, sum)(sum)
```

is allowed now. Sharing between parameters restricts how they can be accessed (concurrently) but it allows the passing of "overlapping" arguments. Extending causality analysis to respect the sharing annotations is straightforward.

9 Implementing Mode Progressions as Synchronous Control Flow

In the previous sections, we have introduced the *Blech* language by means of a PID controller example and subsequently discussed the language's design regarding causality and separate compilation. Now, we turn our attention to how a typical application may be developed using *Blech*. Starting with a rough sketch of the intended behaviour, we design the application step by step while separating concerns and arrive at a concrete implementation. We point out how causality and separate compilation—as defined for *Blech*—help to write easily modifiable components, which can be independently compiled, tested, integrated and refined. We hope to show, how the expressiveness of synchronous control flow is a suitable and maintainable way of implementing the modes of a system that changes its behaviour with the progress of time. The stopwatch example in this section is based on the description by Aguado et al. [1, p. 9ff.]. Subsequently, we modify it and add features bringing it closer to the original presentation in Berry's early work [6].

Imagine a simple stopwatch for counting seconds and minutes. It has two buttons: *StartStop* and *Reset*. The *Reset* always stops the stopwatch and resets it to zero. Pressing *StartStop* the first time starts the stopwatch. Pressing *StartStop* again stops the stopwatch. To resume the measurement, *StartStop* is pressed once again. If both buttons are pressed simultaneously, the stopwatch stops, resets to zero and starts a new measurement all at once. We assume that the stopwatch is running as a timetriggered system where the system tick triggers a reaction step every second.

This mode behaviour can roughly be sketched in a simple state diagram (ignoring the possible simultaneous events).

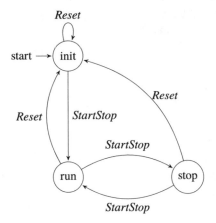

We only use the state diagram as a means of visualisation to assist the problem analysis and to guide the implementation. All the details of the implementation, however, are in the *Blech* code and take advantage of the language's structuring and abstraction mechanisms. This is unlike tools and languages oriented towards graphical, model-based development, which promise graphical programming and code generation as a relief from actual programming. We believe our approach out-matches the annotation of implementation details in purely graphical programming or—even worse—the completion of code in generated code stubs.

Here, the diagram only sketches the modes and the events that trigger a mode change. It does not correctly capture events that happen simultaneously and does not describe how the stopwatch behaves in every mode. Furthermore, the diagram is cyclic and does not naturally imply a sequential flow of mode successions.

Separating concerns, we start by considering the most straightforward use case of a stopwatch: a simple measurement. This brings a natural order to the modes. From the initial mode, we can start the stopwatch, and, after it ran for some time, stop it.

A corresponding sequential synchronous control flow is rather simple. We initialize a counter for the display and wait until *StartStop* has been pressed. While the stopwatch is running, we react to system ticks by incrementing a counter. According to the synchronous semantics, *StartStop* can only be pressed once in every time step, which implies an await between every mode change.

```
activity StopWatchController (isPressedStartStop: bool)
                             (display: Display)
    // init
    display:resetToZero()                                           4
    await isPressedStartStop
    // run
    repeat
        await true
        display:increment()                                         9
    until isPressedStartStop end
    // stop
end
```

The careful reader may object that *isPressedStartStop* could more naturally be declared as a (possibly void) event. However, this is an implementation detail that is up to the designer of the runtime environment which triggers our program and we assume simple boolean flags are used in this case.

Building upon the previous use case, we add the next requirement, namely that the stopwatch may resume measurement. Thus, we allow to alternate between the modes *run* and *stop* by pressing *StartStop*.

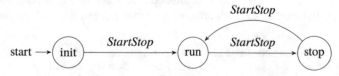

In order to implement this, we add a loop around modes *run* and *stop*. This corresponds to the transition back to *run*. The additional **await** statement in line 13 is the transition guard.

```
activity StopWatchController (isPressedStartStop: bool)
                             (display: Display)
    // init
    display:resetToZero()
    await isPressedStartStop                                    5
    repeat
        // run
        repeat
            await true
            display:increment()                                10
        until isPressedStartStop end
        // stop
        await isPressedStartStop
    end
end                                                            15
```

Finally, let us add the remaining requirement that the stopwatch may be reset in every mode. We add the corresponding *Reset* events in the diagram and essentially arrive at the diagram we have shown in the beginning. However, this time we need to clarify what happens when both buttons are pressed at the same time. From the description above, we understand that the reset takes precedence over the starting or stopping of the stopwatch. This is indicated using the priorities in the diagram: in every reaction step, first, if there is a *Reset* event present, take the corresponding transition and then, if the *StartStop* event is present, take that transition too.

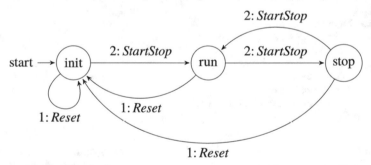

Reinitialising the stopwatch on a *Reset* event can easily be added to our skeleton by using a synchronous preemption. Intuitively, the semantics is as follows: The reset statement marks a position where the control flow restarts when the given condition becomes true before anything else happens at any enclosed *await* statement. Here, we restart the sequential control flow when a *Reset* event occurs before the current enclosed await is evaluated. We wait for a *StartStop* event or we directly start the stopwatch if the *StartStop* event occurred simultaneously (lines 7–9).

```
activity StopWatchController (isPressedStartStop: bool,
                             isPressedReset: bool)
                             (display: Display)
    reset when isPressedReset before
        // init                                                 5
        display:resetToZero()
```

```
        if not isPressedStartStop then
            await isPressedStartStop
        end
        repeat                                                       10
            // run
            repeat
                await true
                display:increment()
            until isPressedStartStop end                             15
            // stop
            await isPressedStartStop
        end
    end
end                                                                  20
```

Finally, consider the complete program that includes the type definition of the *Display* type including its helper functions and the main program that composes the controller with the viewer.

```
/// Display
struct Display
    var seconds: int32
    var minutes: int32
with                                                                 5
    mutating function this:resetToZero ()
        this.seconds = 0
        this.minutes = 0
    end
                                                                     10
    mutating function this:increment ()
        this.seconds = this.seconds + 1
        if 60 == this.seconds then
            this.minutes = this.minutes + 1
            this.seconds = 0                                         15
        end
    end

    /// implemented in C
    @[CFunction(source="display.c")]
    extern function this:show ()                                     20
end

/// Mode progression
activity StopWatchController (isPressedStartStop: bool,              25
                             isPressedReset: bool)
                            (display: Display)
    reset when isPressedReset before
        // init
        display:resetToZero()                                        30
        if not isPressedStartStop then
            await isPressedStartStop
        end
        repeat
            // run                                                   35
```

```
        repeat
            await true
            display:increment()
        until isPressedStartStop end
        // stop                                              40
        await isPressedStartStop
    end
  end
end
                                                             45
/// Main Program
@[EntryPoint]
activity Main (isPressedStartStop: bool, isPressedReset: bool)
    var display: Display
    cobegin                                                  50
        run StopWatchController(isPressedStartStop,
                                isPressedReset)
                                (display)
    with
        // render                                            55
        repeat
            display:show()
            await true
        end
    end                                                      60
end
```

Our *Display* structure is designed to match the data structure in [1]. It therefore has two fields that represent minutes and seconds, respectively. Three methods are attached to the *Display* type: *resetToZero*, *increment* and *show*. The meaning of the first two should be obvious. The last one binds to a platform-specific function which, given a *Display* instance, will actually render the digits on a screen. An annotation gives a hint to the build system where the implementation of *show* can be found. The extern keyword indicates that the *Blech* program will invoke this C function via the foreign function interface. The *Main* activity receives the boolean flags from the calling runtime environment. Additionally, it maintains an instance of the *Display*. In every tick, *Main* will trigger a step in the *StopWatchController* and invoke the *show* function to display the currently measured time.

Note that this simple program exhibits some of the main features that we have discussed in this article. The platform code that runs the application, the controller which we designed in this section, and the platform-specific functionality such as the *show* implementation can all be developed and tested independently as long as they respect fixed APIs. The integration of these separately compilable units happens in *Main*. Causality analysis checks the data flow of the concurrent branches in *Main* and ensures that a reaction of *StopWatchController* is scheduled before the call to *show*.

Also note that unlike the Esterel code in [6], there is no need to externalise all data manipulation to host code. Furthermore, as explained before, our activities, unlike Esterel modules, are composed as black-boxes and causality does not require a global analysis. The programmer reasons locally at the level of one `cobegin` statement and, if any causality issues occur, understands and fixes such issues at this local level.

We have clearly separated concerns within our application. The data management is encapsulated in the *Display* structure, the mode switching is implemented in *StopWatchController* and, finally, *Main* integrates these parts into one runnable app. During implementation we used state transition diagrams as a means to design and document our implementation, without the need for capturing every implementation detail graphically.

Clearly, software is never finished and as such needs to be adapted as customer requirements change or the platform evolves. We discuss two foreseeable changes that go beyond the presentation in [1] but which are taken into account by the original implementation in [6].

First, a stopwatch that runs with a period of 1 s will hardly be useful. Assume we are given an updated platform that ticks every 10 ms which allows us to measure time up to a hundredth of a second. This platform comes with a different screen and an apt implementation of *show* that now consumes a *Display* instance of the following form:[9]

```
struct Display
    var hundredth: int32
    var seconds: int32
    var minutes: int32
with
    mutating function this:resetToZero ()
        this.hundredth = 0
        this.seconds = 0
        this.minutes = 0
    end

    mutating function this:increment ()
        this.hundredth = this.hundredth + 1
        if 100 == this.hundredth then
            this.seconds = this.seconds + 1
            this.hundredth = 0
        end
        if 60 == this.seconds then
            this.minutes = this.minutes + 1
            this.seconds = 0
        end
    end

    @[CFunction(source="display.c")]
    extern function this:show ()
end
```

---

[9]With the need for creating a software library for different platforms, it is obvious that support for interface-based programming is clearly a requirement for Blech.

Nothing in the mode-control or top-level composition changes. With this little change to the data model and its management code, the stopwatch can now run at a higher resolution. It behaves as before.

Now assume the manufacturer wants to use the *Reset* button in the *run* mode to measure lap time. Thus, we rename the button to *ResetLap* and the behaviour is now as follows: Pressing *StartStop* initially starts the stopwatch. Pressing *StartStop* again stops the stopwatch. To resume the measurement, *StartStop* is pressed once again. Pressing *ResetLap* while the stopwatch is idling resets it to zero. Pressing *ResetLap* while a measurement is running displays the lap time while the stopwatch continues to count ticks in the background. Pressing *ResetLap* again switches the display to the currently running total time measurement. The lap time is the interval between time point 0 and the first time *ResetLap* is pressed during measurement or subsequently the interval between two successive laps. We now deliberately ignore the case that both buttons may be pressed simultaneously. The likelihood to accomplish this within a hundredth of a second is very little and hence this feature could hardly be used reliably. Technically, of course, we could again assign priorities in both the running and idling modes.

The above description leads us to a new state transition diagram where we add a new mode to distinguish whether we need to show the total or the lap time.

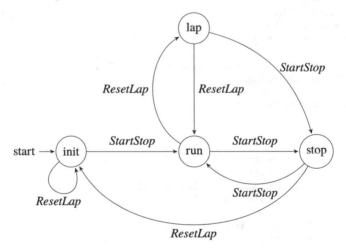

The diagram helps us not only in structuring our modes but also helps us to uncover cases that remain unclear from the textual description above. What happens if the user presses *StartStop* while the lap time is displayed? We *choose* to transition to the idling mode where the *total* time is always shown.

For the implementation, we need to observe that our app has to maintain two timings: the running total time and the point in time where the last lap has been measured to compute the next lap time when needed. Our representation of time is too much oriented towards the human-readable view of a point in time but is not convenient for calculation. Yet, we do not want to break the integration with our *show* function which expects a structure with three fields. Hence, in the following implementation we choose to maintain time in two tick counters *totalTime* and *lastLap* internally in our controller. The function *writeTicksToDisplay* will update the *display* object to the time given by a number of *ticks*.

```
function writeTicksToDisplay (ticks: int32)(display: Display)
    let seconds = ticks / 100
    display.minutes = seconds / 60
    display.seconds = seconds - 60 * display.minutes          4
    display.hundredth = ticks - 100 * seconds
end

activity Measurement (isPressedResetLap: bool)
                     (totalTime: int32, lastLap: int32,       9
                      display: Display)
    // run / lap
    cobegin
        repeat
            await true
            totalTime = totalTime + 1                         14
        end
    with
        repeat
            // show total time every tick                     19
            repeat
                writeTicksToDisplay(totalTime)(display)
                await true
            until isPressedResetLap end
                                                              24
            // calculate lap and update display once
            let lapTime = totalTime - lastLap
            lastLap = totalTime
            writeTicksToDisplay(lapTime)(display)
            await isPressedResetLap                           29
            // back to total time
        end
    end
end
                                                              34
activity StopWatchController (isPressedStartStop: bool,
                             isPressedResetLap: bool)
                             (display: Display)
    var totalTime: int32
    var lastLap: int32                                        39
    repeat
        //init
        totalTime = 0
        lastLap = 0
        writeTicksToDisplay(totalTime)(display)               44
        await isPressedStartStop // transition init -> run
        repeat
            abort when isPressedStartStop after
                run Measurement(isPressedResetLap)
                               (totalTime, lastLap, display)  49
        end
```

```
            // stop, show total time and wait
            writeTicksToDisplay(totalTime)(display)
            await isPressedStartStop or isPressedResetLap
            // back to run if only StartStop was pressed          54
        until isPressedResetLap end
        // back to init if ResetLap was pressed
    end
end
```

When developing the above code, we essentially follow the same steps as before. We consider the sequence of mode switches *init* to *run* to *stop* and the jump back to *run*. This gives rise to lines 41–45 and 51–52. The toggling between *run* and *lap* is factored out into a new activity: *Measurement*. With every reaction, it increments its *totalTime* counter while it concurrently may switch between displaying the total time or calculating and displaying the lap time. Note that the *Measurement* activity only reacts to the *isPressedResetLap* flag. With the synchronous preemption abort in lines 47–50, we mark the end of the block as the position where the control flow continues when the given condition becomes true after any computation in a time step inside the called activity has been executed. That means, it is the calling code that stops the measurement if the *StartStop* button is pressed and proceeds to the next mode *stop*. There, in lines 51–53, the display is set to the total time and the program awaits that either button is pressed. The exit condition in line 55 decides whether we follow the transition to *run* or *init*.

We have shown two possible modifications to our stopwatch implementation. The modifications addressed separate concerns: the first changed the representation of the display and its rendering, the second changed the control logic. Our code reflects this separation and no global changes were necessary. We were able to maintain the APIs and make local changes only.

This example illustrated our approach starting with a specification and arriving at production code as well as subsequent modifications. We believe *Blech* is well suited for embedded application development. It facilitates writing separate units of code which are testable and reusable. Integration and, in particular, concurrent composition is made easy because the reasoning is always done locally. The synchronous control flow of activities enables the programmer to express mode switching logic more intuitively than handcrafted state machines in C which we see in today's production code.

10   Ongoing and Future Work

The *Blech* to C compiler we are currently working on is able to translate all the shown control flow structures and value-types including structs and arrays. Completing the remaining features is an ongoing process. Some of the work presented here forms the necessary prerequisite for an integration of *Blech* and C. It should be possible to call e.g. external C library functions directly from a *Blech* program. This will be a crucial step to enable the use of *Blech* in an industrial context.

Acknowledgements The original version of this article appeared at the FDL 2018 and we thank Prof. Michael Mendler for his invitation to contribute. We also thank the anonymous reviewers as well as Mark Andrew, Jens Brandt, Stephan Scheele, Matthias Terber and Simon Wegendt for their valuable feedback. Finally, we thank Daniel Große, Tom Kazmierski and Sebastian Steinhorst for their invitation to extend and contribute our article to this book.

References

1. Aguado, J., Mendler, M., Pouzet, M., Roop, P. S., & von Hanxleden, R. (2017). Clock-synchronised shared objects for determinisitic concurrency. In *Research Report Bamberger Beiträge zur Wirtschaftsinformatik und Angewandten Informatik*, p. 102. Bamberg: Otto-Friedrich-Universität.
2. Aguado, J., Mendler, M., Pouzet, M., Roop, P. S., & von Hanxleden, R. (2018). Deterministic concurrency: A clock-synchronised shared memory approach. In A. Ahmed (Ed.), *Proceedings of Programming Languages and Systems—27th European Symposium on Programming, ESOP 2018, Held As Part of the European Joint Conferences on Theory and Practice of Software, ETAPS 2018*, Thessaloniki 14–20 April 2018. Lecture notes in computer science (Vol. 10801, pp. 86–113). Berlin: Springer.
3. Ahamad, M., Neiger, G., Burns, J. E., Kohli, P., & Hutto, P. W. (1995). Causal memory: Definitions, implementation, and programming. *Distributed Computing, 9*(1), 37–49.
4. Benveniste, A., Caspi, P., Edwards, S. A., Halbwachs, N., Le Guernic, P., & De Simone, R. (2003). The synchronous languages 12 years later. *Proceedings of the IEEE, 91*(1), 64–83.
5. Benveniste, A., Caillaud, B., & Raclet, J. B. (2012). Application of interface theories to the separate compilation of synchronous programs. In *Proceedings of the 51th IEEE Conference on Decision and Control, CDC 2012*, Maui, 10–13 December 2012 (pp. 7252–7258). Piscataway: IEEE.
6. Berry, G. (1989). *Programming a digital watch in Esterel v3*. Le Chesnay: INRIA. Research Report RR-1032.
7. Brandt, J., & Schneider, K. (2009). Separate compilation for synchronous programs. In Falk, H. (Ed.), *12th International Workshop on Software and Compilers for Embedded Systems, SCOPES '09*, Nice, 23–24 April 2009 (pp. 1–10). New York: ACM.
8. Colaço, J. L., Pagano, B., & Pouzet, M. (2017). SCADE 6: A formal language for embedded critical software development (invited paper). In Mallet, F., Zhang, M., & Madelaine, E. (Eds.), *11th International Symposium on Theoretical Aspects of Software Engineering, TASE 2017*, Sophia Antipolis, 13–15 September 2017 (pp. 1–11). Piscataway: IEEE.
9. Hanxleden, R. V., Mendler, M., Aguado, J., Duderstadt, B., Fuhrmann, I., Motika, C., et al. (2013). *Sequentially constructive concurrency—A conservative extension of the synchronous model of computation*. Kiel: Christian-Albrechts-Universität zu Kiel, Department of Computer Science. Technical Report 1308. ISSN 2192-6247.
10. Hanxleden, R. V., Mendler, M., Aguado, J., Duderstadt, B., Fuhrmann, I., Motika, C., et al. (2014). Sequentially constructive concurrency—A conservative extension of the synchronous model of computation. *ACM Transactions on Embedded Computing Systems, 13*(4s), 144:1–144:26.
11. Lublinerman, R., Szegedy, C., & Tripakis, S. (2009). Modular code generation from synchronous block diagrams: Modularity vs. code size. In Shao, Z., & Pierce, B. C. (Eds.), *Proceedings of the 36th ACM SIGPLAN-SIGACT Symposium on Principles of Programming Languages, POPL 2009*, Savannah, 21–23 January 2009, pp. 78–89. New York: ACM.
12. Potop-Butucaru, D., Edwards, S. A., & Berry, G. (2007). *Compiling esterel*. Berlin: Springer.
13. Sant'Anna, F., Ierusalimschy, R., Rodriguez, N., Rossetto, S., & Branco, A. (2017). The design and implementation of the synchronous language CÉU. *ACM Transactions on Embedded Computing Systems, 16*(4), 98:1–98:26.

14. Schneider, K., & Brandt, J. (2017). *Quartz: A synchronous language for model-based design of reactive embedded systems* (pp. 29–58). Dordrecht: Springer.
15. Von Hanxleden, R., Duderstadt, B., Motika, C., Smyth, S., Mendler, M., Aguado, J., Mercer, S., & O'Brien, O. (2014). Sccharts: Sequentially constructive statecharts for safety-critical applications: Hw/sw-synthesis for a conservative extension of synchronous statecharts. In O'Boyle, M. F. P., & Pingali, K. (Eds.), *ACM SIGPLAN Conference on Programming Language Design and Implementation, PLDI '14*, Edinburgh, 09–11 June 2014 (pp. 372–383). New York: ACM.
16. Wikipedia contributors (2019). *Pid controller—Wikipedia, the free encyclopedia*. https://en.wikipedia.org/wiki/PID_controller. Accessed 2 May 2019.

Index

Printed in the United States
By Bookmasters